8/05

D1176695

The Human Genome Diversity Project

The Human Genome Diversity Project was launched in 1991 by a group of population geneticists whose aim was to map genetic diversity in hundreds of human populations by tracing the similarities and differences between them. It quickly became controversial and was accused of racism and "bad science" because of the special interest paid to sampling cell material from isolated and indigenous populations. The author spent a year carrying out participant observation in two of the laboratories involved in analysis of genetic diversity and provides fascinating insights into the daily routines and technologies used in those laboratories and also into issues of normativity, standardization and naturalization. Drawing on debates and theoretical perspectives from across the social sciences, M'charek explores the relationship between the tools used to produce knowledge and the knowledge thus produced in a way that illuminates the Diversity Project but also contributes to our broader understanding of the contemporary life sciences and their social implications.

AMADE M'CHAREK is Associate Professor at the Department of Biology and the Department of Political Science, University of Amsterdam and is Lecturer in Science, Technology and Public Management.

Cambridge Studies in Society and the Life Sciences

Series Editors
Nikolas Rose, *London School of Economics*
Paul Rabinow, *University of California at Berkeley*

This interdisciplinary series focuses on the social shaping, social meaning and social implications of recent developments in the life sciences, biomedicine and biotechnology. Original research and innovative theoretical work will be placed within a global, multicultural context.

Titles in series

The Human Genome Diversity Project
An Ethnography of Scientific Practice

AMADE M'CHAREK
University of Amsterdam

CAMBRIDGE
UNIVERSITY PRESS

PUBLISHED BY THE PRESS SYNDICATE OF THE UNIVERSITY OF CAMBRIDGE
The Pitt Building, Trumpington Street, Cambridge, United Kingdom

CAMBRIDGE UNIVERSITY PRESS
The Edinburgh Building, Cambridge, CB2 2RU, UK
40 West 20th Street, New York, NY 10011–4211, USA
477 Williamstown Road, Port Melbourne, VIC 3207, Australia
Ruiz de Alarcón 13, 28014 Madrid, Spain
Dock House, The Waterfront, Cape Town 8001, South Africa

http://www.cambridge.org

First published 2005

Printed in the United Kingdom at the University Press, Cambridge

Typeface Times 10/13 pt.   *System* LATEX 2_ε [TB]

A catalog record for this book is available from the British Library

ISBN 0 521 83222 5 hardback
ISBN 0 521 53987 0 paperback

Contents

Preface

Just like any other text, this book embodies many hidden stories. It combines different worlds and is based on the help and effort of many colleagues and friends. Although writing involves solitary journeys, I could not have realised that the secret to writing a book lies in collective work. In a sense, this book produced its networks of intellectual exchange; writing it taught me many things about work and life in academia and in the process it gifted me many friends and colleagues. The book was written, but it also wrote my life.

My interest in genetics existed before I started this project. However, my anxieties and excitement about its potentials came with my work on the Human Genome Diversity Project. Genetics became something that I found myself criticizing or defending, depending on the context that I was in. It thus became my intimate other. I attribute this involvement to the generosity of the scientists who allowed me to take a look in their kitchen and to try out some of the recipes myself. Gert-Jan van Ommen played a crucial role in this. He became involved with my research from the beginning and was a very careful reader of my work, providing me with valuable comments and suggestions. I thank him for long and insightful discussions, and for opening the doors to the community of population geneticists, which enabled me to enter the laboratories. One of the contacts that Gert-Jan helped to establish was with Peter de Knijff, the head of the Forensic Laboratory for DNA Research. Because of the people who work there, the laboratory is a great place to be. I want to thank them all for a very good time, especially Claus van Leeuwen, who took the effort to teach me how to do techniques such as the polymerase chain reaction and DNA sequencing. Peter I thank for taking a social scientist on board, investing all that time and space in my research and for his engagement and lengthy discussions. Since we have continued working together after that period, I am confident that our conversation and collaboration will continue.

Svante Pääbo facilitated my second field study. His Laboratory for Human Genetics and Evolution houses so many talents, and I feel privileged to have been there. I want to thank Svante for providing that space and for the many discussions that we had on topics such as population studies, issues of race and origin. I am indebted to all laboratory members, for it seemed that there was always somebody around whenever I needed help or advice, and especially for being such good company. I feel privileged to have got to know Valentin Börner and Maris Laan and I thank them for their friendship.

Numerous colleagues have contributed by commenting on chapters or versions of the manuscript. I am indebted to them all, even though I will mention just a few. First of all, there is Annemarie Mol. Her insight and ways of looking at the world have become so entangled with mine that I can no longer separate these out in this book. Many of the ideas laid down in the chapters that follow were generated during the long walks that we took. I thank her, not only for her enthusiasm about this project, for encouraging me to complete it, for being such a critical reader commenting thoroughly on the whole manuscript, but especially for her friendship. The many discussions that we had about pressing social and political issues in the world around us in fact made academia a sustainable place to work in and made academic work less of a narrow endeavor. Hans-Jörg Rheinberger's profound knowledge of both genetics and the social studies of science has contributed valuably to this book. I am grateful that he took the time to read a previous version of the manuscript and that he had supplied me with insightful comments and suggestions. I also thank Michael Lynch for careful reading of several chapters and for pertinent suggestions to improve these. Mieke Aerts and I only got to know each other in the recent years. Our mutual interest in each other's work has started an ongoing conversation. It is in discussions with Mieke that I found a way of talking about method. I am grateful for that. My work has also benefited from the insights of John Law, both in his own work and in his comments on my papers. I also thank him for inviting me to various workshops that he had organized – in Lancaster and elsewhere – and all the participants in these workshops for their feedback on the papers I presented. The University of Lancaster (Department of Sociology and the Department of Women Studies) houses many inspiring scholars and is an exceptional intellectual community. I especially want to mention Claudia Castañeda, Maureen McNeil and Lucy Suchman.

Olaf Posselt was involved in my research in various ways. Over the years, he has read my texts at various stages and grew to be one of my best critics. I cannot thank him enough for all the support that he has given me, nor for the help in solving numerous problems.

Turning my research on the Human Genome Diversity Project into a book was part of a journey through different academic institutes. I started out at the

Belle van Zuylen Institute for Gender and Cultural Studies. I am indebted to all my colleagues there, but especially to Marion de Zanger, Sybille Lammes and Catherine Lord, who helped me to see academic work in context and to place it within a much broader frame. The Amsterdam School for Cultural Analyses provided an interdisciplinary environment where I learned more about semiotics and philosophy. My friendship with Frans Willem Korsten is a precious "spin off" of my participation in the seminars of this school. My work has also benefited from the Summer and Winter Schools of the Netherlands Graduate School of Science, Technology and Modern Culture (WTMC), in which I participated first as a Ph.D. student and later as a guest lecturer. I especially thank Paul Wouters, Ruth Benschop and Ruud Hendriks for feedback and suggestions on several chapters. The department of Science Dynamics and the Department of Biology (University of Amsterdam) employed me as a lecturer, involved in developing an MA program for students in the sciences. I especially want to mention Stuart Blume and Leen Dresen and thank them for numerous conversations and collaborations, and for their friendship. Antje van de Does-Bianchi is a great "boss," and her encouragement and comradeship is invaluable. Mirjam Kohinor and Helen Bergman I thank for being such good colleagues and for being a true team in organizing and teaching the MA course. I also thank Mirjam for her help with the Glossary.

Moving partially to the department of Political Science has created new intellectual challenges. How to do politics with DNA, so to speak, and how to go about diversity have become shared topics. I thank John Grin and Maarten Hajer for making this possible. My collaborations with John started before I joined the department and he has commented on several parts of the manuscript. I thank him not just for that but also for being such a good friend. As a member of the Amsterdam School for Social Science Research (ASSR), I find myself among so many talented colleagues. I want to especially thank Anita Hardon for her enthusiasm and encouragement, also seen in our shared project on diversity in medical practice. I am indebted to the ASSR and especially José Komen for financial support and for the infrastructure for finishing this book.

While finishing this manuscript, Dick Willems, Marjolein Kuijper, Victor Toom, Nicolien Wieringa and I were conducting a joint project on the politics of everyday medical technologies. The different issue that we were discussing in the context of this project have helped me to solve some of the problems that I was dealing with in the manuscript.

Nicholas Rose and Paul Rabinow, the editors of the *Cambridge Studies in Society and the Life Sciences*, invited me to submit my research for publication in this series. I am deeply indebted to them. I thank Nicholas also for detailed comment on each and every chapter. The manuscript has also benefited from comments of three anonymous reviewers. I want to thank them also

1
Introduction

The researcher in the field

On December 15 1996, I went to Munich Airport to meet a professor in population genetics. She had travelled from Tel Aviv to visit the laboratory where I was conducting my research. After we had found each other in the crowd, we took the train back into the city. Professor B.-T. turned out to be a very pleasant person and quite soon we found ourselves in animated conversation. She told me about the rare DNA samples that she had brought along and where she had collected them. The members of the laboratory were looking forward to the samples, specifically because they were running short of male samples from these populations. She had heard that I too was going to use the samples for my research project. I told her about my study and what I had uncovered so far. At the same time, I started to feel a little uncomfortable. I felt the urge to "reveal" my "identity" to her. Because I was not just a member of the laboratory: I was also studying it. However, before I could do so, Professor B.-T. was eager to learn where I came from. I told her that I lived in Amsterdam but that I was originally from Tunisia. A little shy but curious, she asked me whether I was also from "one of those interesting populations." I had to disappoint her there, but I told her about the genealogical history of my family, which dates back over a couple of hundred years and goes back into Lebanon.

Two years later I was visiting Professor B.-T. in Tel Aviv. She invited me to her laboratory and introduced me to her group. I learned that her laboratory housed one of the consortia of the Human Genome Diversity Project (referred to in this book as the Diversity Project) where they were growing cell lines of various population samples. In addition, when she introduced me to her colleagues, I was surprised that I was introduced not as a social scientist from Amsterdam but as a member of the laboratory in Munich.

The stakes and the argument

In 1991, a group of population geneticists embarked on an international project designed to map human genetic diversity.[1] The initiators of the Diversity Project were interested not only in mapping contemporary genetic diversity as such but also in studying how the current diversity had evolved, and how people and genes had spread over the world. Knowledge of the origins of populations, as one of the initiators of the project has stated, would have "enormous potential for illuminating our understanding of human history and identity."[2] By tracing similarities and differences in the DNA of various groups of people, geneticists hoped to reconstruct where humans came from, along which routes they migrated and when, and how different groups of people relate to one another. To do so, a special emphasis is placed on the study of "indigenous peoples" and "isolated populations." They are deemed the "treasure keepers" of original information, which, in the course of history, had gradually been obscured in other large groups because of migration and admixture. Isolated populations are held to be conservative in this respect by geneticists.[3] As distinct populations, their DNA is considered to be representative of human genetic diversity at large and, therefore, convenient for attaining the goals of the Diversity Project.

The Diversity Project was launched with rhetoric of preservation, time pressure and alarm. In June 1991 the journal *Science* published an article entitled *A Genetic Survey of Vanishing Peoples*, which opened: "Racing the clock, two leaders in genetics and evolution are calling for an urgent effort to collect DNA from rapidly disappearing populations."[4] One of them, the Stanford population geneticist Luca Cavalli-Sforza, argued that "if sampling is too long delayed, some human groups may disappear as discrete populations. . . . At a time when we are increasingly concerned with preserving information about diversity of the many species with which we share the Earth, surely we cannot ignore the diversity of our own species."[5]

However, the Diversity Project soon ran into trouble. It was faced with a variety of criticisms, especially from indigenous and environmental organizations. It was soon dubbed "The Vampire Project," referring to the collecting of blood samples.[6] Furthermore, this naming seemed to suggest that the people sampled were ill-informed and misled by geneticists and that the samples were collected for interests other than those of the sampled groups. In the television documentary *The Gene Hunters*, the Professor of Medical Ethics at Massachusetts Institute of Technology George Annas said: "We're taking from them their DNA, which we now consider like gold. It's even worse than standard colonialism and exploitation, because we are taking the one thing that *we* value,

and after we take that, we have no real interest in whether they live or die."
In that same documentary, the spokesperson for the Arhuaco People, Leonora
Zalabata, stated: "Our land, our culture, our subsoil, our ideology, and our tradi-
tions have all been exploited. This [the Diversity Project] could be another form
of exploitation. Only this time, they are using *us* as raw material."[7] The criticism led to heated debates about the social and ethical aspects of the Diversity
Project. In 1993, the Rural Advancement Foundation International (RAFI) as
well as other political agents urged geneticists to incorporate indigenous organizations in every step of the Diversity Project and to reassess its scientific and
ethical implications. By the mid-1990s, many other organizations, including
the Bioethics Committee of UNESCO and the US National Research Council
(NRC), were calling for strict regulations on how to sample and handle the
information obtained.[8] The project had also become part of a debate about
commercial revenues in science, such as the patenting of human genes and the
development of drugs for specific diseases.[9] Geneticists, however, emphasized
that their initiative had no commercial interests, nor would they accept funding
from commercial agents.[10] They argued that the knowledge resulting from the
Diversity Project might contribute to the understanding of genetically inherited
diseases but its major goal was an investigation of genetic diversity and the
history of human migration. This "pure science" approach was looked at with
suspicion, for example by Ray Apodaca, a spokesman of the National Congress
of American Indians. Countering the "pure science" claims he stated: "We know
where we came from, and we know who we are, and we think we know where
we are going. Why do we need to know anything else? I mean, is this for their
benefit? It certainly isn't for ours."[11]

In the face of this criticism, the Diversity Project met initial problems finding
financial or other support within the scientific community and institutions.[12] In
Europe, the Human Genome Organization (HUGO) proved at an early stage
to be willing to finance a series of workshops in order to assess the project's
scientific values. Some US organizations, such as the US Science Foundation
and the US Department of Energy, followed this by funding three planning
meetings.[13] However, whereas the European Union had supplied 1.2 million
dollars to set up laboratories where European genetic diversity could be studied, the project was put on hold for several years in the USA. By the end
of 1997, however, a committee of the US NRC had evaluated the project
and found that it should receive financial support within American national
borders, provided that it met ethical and legal restrictions placed on genetic
research funded by federal agents.[14] While few research projects received
financial support, a number of diversity consortia for the storage of samples
and the growing of cell lines were established, such as the one I encountered

in Tel Aviv. Thus, although haltingly and dispersed, the Diversity Project started.

This book is about the Diversity Project. More specifically, it deals with genetic diversity in scientific practice. Prompted by the issue of "conserved genes" and the mapping of similarities and differences between populations, it focuses on what genetic diversity is made to be in scientific practice. This brief review of the controversy shows some of the political stakes in the Diversity Project. Rather than a study of that controversy and of the different politics involved in the debate outlined above – however important and interesting in its own right – this book aims at tracing the politics of genetic diversity in laboratory routines. It investigates the daily practice in which humans, samples and technology are aligned to produce the stuff of which the power and prestige of science is made. The argument pursued throughout this book is that genetic diversity is not an object that lies waiting for the scientist to discover, nor can it be treated as a construct of scientists. Genetic diversity is enacted in a complex scientific practice. It is not only dependent on the scientist and the DNA but also on the various technologies applied to produce it.[15]

Let me briefly illustrate the relevance of technologies for the Diversity Project. For instance, the haste with which geneticists attempted to "conserve" human diversity before "isolated populations" ceased to exist as such cannot be explained exclusively in socio-cultural terms, or as a sudden interest in (bio)diversity. What is at stake is not the fact that the lives of these groups of people are endangered or that their integrity is threatened because they nowadays tend to migrate and mix more frequently with other groups than in previous times – a so-called "death by reproduction" as Corine Hayden[16] aptly termed it – nor is it that these groups only came to the attention of geneticists in the late 1980s. Many of the geneticists participating in the Diversity Project had already been studying and comparing these populations previously and some had even stopped doing so in the 1970s because they "ran out of data."[17] With the technology available at that time, these scientists could acquire no more information from the samples they had. What *did* change by the end of the 1980s was the availability of new technologies. The introduction of revolutionary technologies to the field of genetics had made it possible to produce new "data" based on the samples already collected and also brought within reach a study of diversity on a much larger scale. What these technologies are and how they affect genetic diversity is, therefore, at the center of this book. Consequently, rather than *whether or not in our genes*,[18] the question addressed is *how* in *whose* genes? Before going into the details and the organization of this book, let us first go back to the Diversity Project to have a second look at how it is (intended to be) organized.

The Human Genome Diversity Project

The Diversity Project did not emerge in isolation. Many more genome projects were launched in the 1990s and before.[19] Most powerful is the *Human Genome Project* (HGP), aimed at *the* human genome, which had been presented to the world in June 2000 by science, commerce and politics.[20] Since the Diversity Project was announced by its initiators as a response to the HGP, I will elaborate on this. The goal of the HGP was a map of the complete human genome.[21] The sequence map would function as a reference genome against which all human individuals could be located and compared. As *the* reference, it would provide the genetic terms in which all individuals would be expressed.[22] One of the initiators, the geneticist Walter Gilbert, presented the HGP as the ultimate means to know oneself. He insisted most strongly that molecular biologists would have the final answer to what it is that makes us human, namely the DNA. One of his frequently quoted statements was that: "one will be able to pull a CD out of one's pocket and say, 'Here is a human being; it's me!'"[23]

The compact disc (CD) metaphor is obviously a pregnant one, not only because it allowed Gilbert to make his argument tangible during his presentations by actually pulling a CD out of his pocket but also because it underlined the technical aspects of genomes and genetics. However, riding on that metaphor, the political stakes are not only in knowing the information it contains but also in how and where the CD is produced. What kinds of polymerized substance, stencil-plate and printing technologies contribute to the CD? How can it be played and what kind of equipment is necessary? How can it be read and who will be able to read it? Who will have a CD? What about the possibility of copying it? Will the result be a copy or a clone? Also, what kind of place will the CD-of-life take in the collections of those who have many different CDs? Will it be able to compete with a CD containing a family photo album, with one bearing a game called *Doom* or with that of a singer called *Fairouz*? What kind of practices make the one CD more important than the other? These questions encompass more and more people and things and make them and the relations between them part and parcel of the CD-of-life. In addition, since the goal of the HGP is to produce *one* CD, a question raised within and outside the confines of genetics is, whose CD is it going to be?

> The first complete human sequence was expected to be that of a composite person: it would have both an X and a Y sex chromosome, which will formally make it a male, but this "he" would comprise autosomes [non-sex chromosomes] taken from men and women of several nations – the United States, the European countries, and Japan. He would be a multinational and multiracial melange, a kind of Adam II, his encoded essence revealed for the twenty-first century and beyond.[24]

This was written by Daniel Kevles, half ironically, in *The Code of Codes*, a now classical edited volume about the HGP. However some geneticists outside the realm of the HGP claimed that "[t]he Human Genome Project aims to sequence 'the' human genome with DNA taken mainly from individuals likely to be of European ancestry in North America and Europe. But, like all brothers and sisters, all humans have slightly different genomes."[25] They, therefore, suggested another genome project, the Diversity Project, which "wants to explore the full range of genome diversity within the human family."[26]

Studies of human genetic diversity are not new and go back to the beginning of the twentieth century, when they were based on blood groups. In addition, DNA-based research flourished from the mid-1970s onwards.[27] Hence the initiation of the Diversity Project followed from ongoing research. Yet every project has a myth of origin:[28] there is a date of birth and there are great men involved; there is a vision and there are allies inside and outside the field; there is a world to be gained and ghosts to be exorcized. What follows is the origin myth of the Diversity Project.

The Diversity Project was initiated in 1991 by a group of American geneticists among whom were the late Allan Wilson (Professor of Biochemistry at Berkeley University) and Luigi Luca Cavalli-Sforza (Professor of Population Genetics at Stanford University). Together they found more colleagues welcoming their plan to map genetic diversity of human populations on a worldwide basis.[29] The values of this initiative (referred to in the quote as the HGD Project) were summarized as follows.

> The main value of the HGD Project lies in its enormous potential for illuminating our understanding of human history and identity
>   The resource created by the HGD Project will also provide valuable information on the role played by genetic factors in predisposition or resistance to disease
>   The HGD Project will bring together people from many countries and disciplines. The work of geneticists will be linked in an unprecedented way with that of anthropologists, archaeologists, biologists, linguists and historians, creating a unique bridge between science and the humanities
>   By leading to a greater understanding of the nature of differences between individuals and between human populations, the HGD Project will help to combat the widespread popular fear and ignorance of human genetics and will make a significant contribution to the elimination of racism.[30]

A central question of population genetics is how did humans migrate out of Africa to 'colonize' other regions in the world, and when did these events take place.[31] The idea is that human genetic makeup is indicative of historical events and vice versa: that the contingency of human history is reflected in the DNA. By tracing similarities and differences in the DNA fragments of various populations, geneticists hoped to provide another (a better?) account of human history. Culture and nature are thus married-up in the Diversity Project.

> There is a cultural imperative for us to respond to that opportunity and use the extraordinary scientific power that has been created through the development of DNA technology to generate – for the benefit of all people – information about the history and evolution of our own species.[32]

To reach this goal the initiators aspired to create an internationally organized project: a project based on technologies and knowledge developed within the realm of the HGP and capable of redirecting the work conducted in the field of population genetics. As early as 1991, the Diversity Project was "adopted" by HUGO, which had been established in 1989 within the HGP. To assess the potentials of the project in Europe, HUGO set up an ad hoc committee in the autumn of 1991. This committee was charged with organizing a series of workshops where various aspects of the project were to be discussed and evaluated, such as the methods of sampling and the storage of the samples, the technologies to be applied and the processing of the information, and the social and ethical aspects of the project. The committee was also asked to conduct a pilot study, using already existing samples, to establish the relevance and added value of the project and to adjust the protocols for the forthcoming research.[33] In the first five years, the project as a whole was estimated to cost 25–30 million American dollars. HUGO provided 1.2 million dollars to organize the workshops and to conduct a pilot study. Additionally, HUGO helped to create a friendlier political climate for the project to get started. Parallel to this, a number of agencies in the USA had supplied some funding for the organization of three workshops to deal with the sampling strategy, the selection of the populations to be studied and the technologies and ethics of the project. The Diversity Project is now organized in a number of regional committees responsible for their own initiatives.[34] Whereas the European regional committee was receiving European Union support as early as 1992, the North American regional committee had to wait until 1997 for federal support.[35]

Making a genetic map of the world

A major goal of the Diversity Project is to make a map of the world that shows genetic relief and contours. Such a map would reconstruct human migration out of Africa and the spread of humans and their genes around the world; the intention was to be able to assign different *populations* to different loci on that map. Yet its two initiators, Cavalli-Sforza and Wilson, already had conflicting ideas about the sampling strategy; that is, about what a *population* is. Whereas Cavalli-Sforza had strong ideas about how to define a population, namely on the basis of linguistic criteria, Wilson argued against any presupposition about what population is. In an interview with *Science*, Wilson stated: "We should abandon previous concepts of what populations are and go by geography. We need to be explorers, finding out what is there, rather than presuming we know what a population is." Hence his idea was that what population *is* should be the outcome of genetic research and not the start. He, therefore, suggested a grid sampling based on geographical distances (100 miles).[36] The grid approach, however, was considered too costly in terms of time and money, and categorization according to linguistic criteria was regarded as the most appropriate.[37]

In addition, in the NRC evaluation of the Diversity Project, the term population was bracketed.

> The term "population" has many meanings; it is most often used to designate a body of persons (or other organisms) that have a common quality or *characteristic*, to designate a group of interbreeding organisms, or to designate a group of persons (or other organisms) that occupy a specific geographical locale.[38]

Taking linguistic criteria as characteristics, geneticists were faced with 5000 different populations.[39] Yet, as in the case of a geographical grid, sampling, storing and studying the cell material of all these groups did not seem feasible. The initiators, therefore, decided to focus on just 500 populations. The selected populations should do the following.

> . . . answer specific questions about the processes that have had a major influence on the composition of current ethnic groups, language groups, and cultures . . . [This suggests a study of] populations that are anthropologically unique; populations that constitute linguistic isolates; populations that might be especially informative in identifying the genetic etiology of important diseases; and populations that are in danger of losing their identity as recognizably separate cultural, linguistic, or geographic groups of individuals.[40]

These qualities not only give clues about what it means to be genetically representative or to enable the tackling of "interesting" questions. They also suggest that the linguistic criterion is highly invested with various notions about the social, the cultural and the biological. In an article published in the *Scientific American*, Cavalli-Sforza reported on the correspondence between the distribution of genes and that of languages among populations. Elaborating on the transmission of genes, language and culture from one generation to the other, he distinguished between a vertical and a horizontal transmission, the first being a transmission between parents and offspring, and the latter a transmission between unrelated individuals. Whereas genes can only be transmitted vertically, culture and language may be passed on by either path. While identifying the difference between "isolated populations" and populations that have undergone admixture, he stated:

> In the modern world horizontal transmission is becoming increasingly important. But traditional societies are so called precisely because they retain their cultures – and usually their languages – from one generation to the next. Their predominantly vertical transmission of culture most probably makes them more conservative.[41]

Hence language is not just an arbitrary means of distinguishing between groups of people: it is deemed to correlate with the genes. More specifically, this correlation is held to be even more elegant when applied to the Diversity Project's main object of study, namely "isolated populations." By analyzing and comparing the similarities and differences found in various of these populations, geneticists hope to gain insight into "genetically complex" populations: populations that are less isolated, less unique and less easy to categorize and to study. It seems that those who are not considered to be connected to the global traffic of humans and things, especially those in far-off places, are considered best sources for understanding how genetic "melting pots" must have come about.[42] Based on the idea that genetic diversity (just like language and culture) is better preserved in "isolated populations" and the idea that all humans belong to one "genealogical family" originating from Africa, these populations are assigned the role of origin and resource.[43] They are thus considered to be more homogeneous and their genetic makeup to be more conserved. However, how can they represent an overall human diversity, such as aimed at by the Diversity Project? In addition to their homogeneity and conserved genes, the genetic makeup of different "isolates" in different parts of the world is held to represent specific moments in the history of human migration. These migration events may also be represented in intermixed groups but their effect on the clustering of genes tends to be blurred through population admixture. This indicates that representing human genetic diversity at large can only be done if different "isolated populations" from different parts of the world are taken into account.

The emphasis placed on "isolated populations" is relevant for studies of diversity not only in the context of human history but also in that of genetic diseases. In a document issued by the Diversity Project, this relevance is phrased as follows: "Every time we ask whether a particular genetic marker is associated with a disease, we need to know about the normal control population. The need for this comparison increases with the diversity of the population."[44] Therefore, in order to understand the mechanisms of inherited diseases in genetically diverse populations, "isolated populations" may function as normal control populations. With the help of such information, geneticists hope to trace where and when specific mutations occurred and whether they lead to the same effects: that is, they also cause diseases in the control population. However, where the specific genes related to a disorder are not known, the role of an "isolated population" might be different. For example, if such a population is susceptible to a specific disease, studying that particular population and not one where genetic diversity is larger may be seen as an application of the reductionist method of the natural sciences.[45] Applied to an object of research, this method consists of reducing complexity to a small number of controllable variables that *can* be studied in a laboratory context. In line with this, "isolated populations" rather than other control groups would function as resource material.[46] As a geneticist once explained to me: "It would be crude to place a wall around Friesland [a province in the Netherlands] and observe what happens to its inhabitants, made isolated. These populations live isolated by nature and can give us insight into the development of various diseases." Although geneticists would consider these populations interesting for studies in their own right,[47] within the context of the Diversity Project they occupy the position of resource and can be seen as a "natural" laboratory for the rest. Whether the aim is to reconstruct the migration history of humans, to preserve (knowledge about) human genetic diversity or to study human genetic diseases, the Diversity Project makes some populations into a more appropriate resource than others.

Studying genetic diversity within the context of a project does not just affect what may be considered a population, what a population is and how it is deemed to contribute to its research, it also affects genetics as a field. Within the Diversity Project, geneticists had to decide upon how to sample, how to store the samples and what kinds of technology will be used to study the samples. To create a project, they simply had to work together and standardization is an important condition for achieving that.

The Diversity Project intended to collect 10 000–100 000 samples from the 500 populations under study. The sampling was delegated to the regional committees, who were asked, where possible, to work together with "local" scientists and anthropologists in the field.[48] When the samples left these regions, they were not to travel alone: they should be accompanied by information about the region and about the sampled individual. The samples were to be accompanied to central storage areas by information regarding "sex, age (or approximate year of birth), current residence, place of birth, linguistic affiliation [of these individuals and] current residence, place of birth, cultural affiliation, linguistic affiliation [of the individual's] biological parents."[49] Thus, the study of the diversity of these populations had to involve more than cell material or DNA. For, as the NRC committee had recommended, "the inclusion of parental birthplaces with the other information identified above could, in some instances, inadvertently reveal a particular person's identity."[50] Identity is to be understood in terms of ethnicity and origin.

From most individuals, only a small quantity of cell material was to be collected: blood, hair root or inner cheek tissue. These samples were to be stored as DNA in DNA libraries. Given the availability of DNA copying technologies, even small quantities of DNA were deemed sufficient for study purposes. However, since some samples were also to be used to produce cell lines, the in vitro growing of cells, more cell material was needed from 10% of the sampled individuals. Their white blood cells would provide the Diversity Project with an endless source of DNA.[51]

In the Diversity Project it was emphasized that the proposed research was not new. It was stated that:

> ... what is new is the possibility of extending the study of population to a much more detailed level by applying some of the DNA technology (such as the PCR-based technology mentioned above) that has been developed within the last few years in the context of the Human Genome Project.[52]

However, to study DNA and thus to know a population, geneticists have different tools at their disposal. Studying a population in terms of height, for instance, by measuring from head to toe does not make that population comparable to another studied in terms of weight, measured in kilograms. Hence one of the major efforts of the Diversity Project in this respect was to coordinate and fine tune the technologies to be applied, such as the kind of DNA copying technologies (e.g. the polymerase chain reaction (PCR)[53]), the specific fragments of variable DNA to be studied (genetic markers) and the statistical models to analyze and compare the data.

As in the HGP, technology is at the center of the Diversity Project.[54] It accounts for the project's potential for population studies. It is argued that "... as a result [of revolutionary technology], the precision with which populations, their origins and their interrelations can be defined, using relatively small samples, has increased enormously."[55] Still, even though the technology is "cutting edge" and allows for genetic studies on the basis of small samples, geneticists find themselves confronted with a problem. "[T]he human species is moving towards increasingly intensive amalgamation" and populations are losing their identities in terms of genetic homogeneity.[56] This is considered to be the "irony" of the Diversity Project. An irony that makes it convenient to study isolated and aboriginal populations instead.

Objections to objectification

In the course of the Diversity Project, there emerged yet another irony. This irony was again related to the project's object of research. Its object turned out to be a subject as well. As Donna Haraway has noted for the populations to be investigated ". . . diversity was both about their object status *and* their subject status."[57] Quite soon, the internationally organized project created opposition on an international level and was met with harsh criticism from the very populations that appeared on its priority lists.[58] On 8th July 1993, the Third World Network placed a call on the list-server of Native-L (an aboriginal First Peoples news net) aimed at stopping the Diversity Project.[59]

> In the year of indigenous people and at the time of UN Conference on Human Rights we find such initiatives emerging from the West totally unethical and a moral outrage. We call on all groups and individuals concerned with indigenous peoples' rights to mobilise public opinion against the case of human communities as material for scientific experimentation and patenting. Indigenous communities are not just "isolates of historical interest". They have a right to be recognised as fully human communities with full human rights which include decision about how other countries will relate to them.[60]

The Third World Network was alerted to the Diversity Project by RAFI. A few months previous, RAFI had expressed its belief of the Diversity Project's threat to the autonomy and livelihood of indigenous groups. In that letter, addressed to the organizers of the Diversity Project, they also argued that this initiative might divert money from basic services, such as medication and clean water, to the compilation of gene banks.[61] Following this call, a debate ensued on Native-L in which some of the project's organizers participated, and quite soon this debate went beyond the borders of this news net.[62] Tribal governments and other organizations of peoples around the world started to create problems for the project. Whereas the proponents of the Diversity Project emphasized its mere scientific goals and the fact that it would refrain from economic exploitation of the genes and knowledge acquired, in the debates the project was constructed in a different manner. In these debates, the Diversity Project was linked with concerns about economic exploitation in terms of gene patenting and the development of expensive medication, with the development of ethnic weapons ('gene-bombs') and racism, with bio-piracy and bio-colonialism and with a sheer interest in the history of populations rather than their futures. It seems that, however hard the projecteers tried to purify their initiative, it became a focal point for various different concerns that indigenous peoples and others have with biotechnology and the new genetics. Although some populations decided to collaborate in order to learn more about certain diseases that prevailed in their population, or to benefit from the promised technology transfer, many more organized themselves on an international basis against the appropriation of their body tissue. Moreover, even though RAFI as well as the World Council of Indigenous Peoples urged the geneticists involved to organize a meeting with representatives of indigenous peoples to address the ethical issues of the Diversity Project, their request has not yet been met.

As well as being dubbed the "Vampire Project" by the World Council of Indigenous Peoples, the Diversity Project was also described as "bad science," a synonym for racist science in the period after the Second World War.[63] The joint interest in genes and populations was considered to reify biological races and to essentialize differences. Despite the promise of amending arrived ideas about biological difference between the races, or populations for that matter, in the debates about the Diversity Project many divides are being reified, such as the melting pot in the West versus the isolate in far-off places, and the genetically heterogenous here versus the genetically pure there. It is a separation between the world of science and technology and the world of the resourced object of study.[64] In this divide, it appears as if (isolated) populations are out there, in nature, waiting for geneticists and their technologies to reveal their histories. However, the idea that human history can be read from the genes is problematic, not only because it tends to ignore already existing accounts of history and origin.[65] It also raises the question of what makes some populations into "isolates" or "indigenous" and not others. The Diversity Project has put more than 500 so-called "isolated populations" on its priority lists. As many anthropologists have shown, the fact that some groups live relatively isolated lives does not mean that it has always been that way.

> The San peoples of South Africa, for example, at the top of the so called 'genetic isolate list', and therefore a pristine example of an uncontaminated population by HGDP standards, embrace three different language groups, suggesting relatively recent formation as a single group . . . the San became isolated only in the 19th century, and their isolation is related directly to colonialism.[66]

What holds for the San people holds also for many other groups, the so-called "people without history."[67] Consequently, genetic isolatedness of populations might as well be an effect of the Diversity Project, which is based on a rather naïve version of history. Those living in far-off places are *re-produced* as genetically (and culturally) homogeneous and pure, and as the genealogical origin of some other groups who have migrated out of the tribe long ago to work their way through the modernist melting pot. Even though the intended contribution to the elimination of racism might be well intentioned, such reductionism of the complex histories of groups of people around the world may indeed contribute to the naturalization of population and the reintroduction of a discourse of race and racism to genetics.[68]

Meanwhile, and as a response to this varied critique, the North American Committee of the Diversity Project has developed an ethical protocol.[69] This model ethical protocol indicated that, should any monetary revenue come out of analyses conducted by diversity researchers, individuals or populations who donated samples for that particular research should be granted monetary compensation. It also introduced the concept of community informed consent next to the individual consent that needed be sought. Therefore, before any sample could be taken from a group of individuals, the community's consent would have to be obtained as well. In this, the protocol referred to the population-based nature of diversity research. Moreover, in cases of potential patents or other marketable products, both the consent of the individual and that of the community would be required, and a respected international organization would be approached to mediate in the negotiations. Subsequent to the development of the protocol, the Diversity Project seems to be moving away from an emphasis on "isolated populations" to one that includes various different groups. European populations are such a case.

Making a book

As can be seen, the Diversity Project is complex, broad and controversial. This increases the ways in which it could be studied.[70] What comes to the fore is its controversial character: its blunt "science for the West and genes from the rest" kind of appearance. While this is disturbingly important, I have chosen a different angle. Instead of contrasting "genes" to "science," in a kind of naturalized dichotomy between nature and knowledge, and instead of a geographical separation between the *worlds* of the populations studied and the *words* of the scientist studying them, my aim was to investigate how they are made into constituent parts of genetic diversity. Where to do my study was a matter of "choice." As I indicated at the beginning of this chapter, I did not choose to study the public debate around the Diversity Project, but nor did I choose to study the populations targeted by this project. Setting out to study such populations because of an interest in the Diversity Project seemed to me invasive. In addition, even though it was easy to side with the criticism against the Diversity Project, it seemed to me that the debate was too neatly organized along the lines of wrong and right or good and bad. This increased my curiosity about the Diversity Project and raised several questions. What is it about? How does it or will it change our worlds? I contend that genetic diversity cannot simply be the end-product of knowledge applied to populations or their DNA:[71] I had become interested in what it involves in scientific practice. Consequently, I turned to laboratories and studied their practices.

As discussed briefly above, the study of diversity requires a certain standardization of practices: hence the emphasis placed on fine-tuning the project's "materials and methods." As pointed out, standardization had to be achieved for "population": how to define populations and how to sample them. For technologies such as thosed used in DNA amplification, it involved the fragments of DNA to be studied as well as the statistical models to be applied.[72] If data on genetic diversity and populations were to be comparable between laboratories, scientific conduct needed to acquire a routine. It is this very routine and the "nothing strange going on here" kind of practice that I examine in this book. Genetic diversity will be traced in such practices. Practices where various technologies are employed routinely to produce it. The reader who is immediately interested in how I do this (i.e. the methods applied) is referred to Chapter 6, where I discuss the issue of method.

The examinations conducted in the next four chapters are guided by the questions: What is genetic diversity and how is it *enacted* in laboratory practice?

Genetic diversity as well as other objects of genetics, is enacted or performed rather than discovered, analyzed or animated. This is one central point of departure of this book. The notions of enactment or perfomativity are used to emphasize that objects (such as diversity, population, individual, etc.) emerge in practices consisting of individuals, technologies, language and theories among others. Objects are dependent on such practices and may fail to exist outside of these. They may be stable or coherent in one practice but not in another, or, it might require extra work to make them move along practices while retaining a stable form.[73] Applying these notions also underlines that there is nothing essentially there in nature or the DNA to be captured in one final form. Rather, as will be shown in the following chapters, different technologies produce different versions of any object.

The four chapters are a collage. As in a collage, they show overlaps between technologies, scientists, scientific publications, laboratory practice and focuses of analyses. As in a collage, some pieces are cut out in order to focus more on others. It is not the aim of this book to map all the different ways that genetic diversity is established, or all the technologies involved in achieving it, not even in the laboratories studied. The aim is to focus on some core practices, technologies and objects in studies of diversity and to examine how they help to produce genetic similarities and differences. Given the research that is being conducted in the Diversity Project, the cases analyzed here are, therefore, simultaneously narrower and broader than the scope of this project. Narrower because they do not take into account all the actors involved in producing genetic diversity. Broader, because the technologies addressed also have relevance for other fields inside and outside the field of genetics.

Each chapter highlights a different aspect of genetic diversity by addressing another practice of making similarities and differences. The chapters can be read in any order. The order I have chosen makes my own narrative of genetic diversity, namely that of standardization, naturalization and diversity.

Chapter 2 deals with *population*. In the Diversity Project, population is *defined* according to linguistic criteria. In this chapter, I examine *practices* and analyze what population is made to be in daily laboratory work. Chapter 3 investigates *genetic markers* (variable DNA fragments) and views how they embody theoretical and methodological notions of diversity. Examining the practicalities of genetic markers in laboratories, this chapter also addresses issues of standardization as envisioned in the Diversity Project. Chapter 4 discusses a mitochondrial DNA *reference sequence*, a piece of technology with which other sequences can be compared. This chapter examines how the reference sequence is applied in laboratories at present but also how it was produced in the 1980s. This case is further analyzed in terms of naturalization and the normative content of technology. Chapter 5 is about *genetic sex* and *genetic lineage*. Here I investigate the various ways in which the sexes are enacted in studies of genetic lineage and show how DNA is treated both as a resource of diversity and as a technology of establishing sexualized lineage. In Chapter 6, the concluding chapter, I take up the narratives about standardization, naturalization and diversity to reflect upon the analyses in the preceding chapters, and their relevance for science and technology studies, genetics, and gender and anti-racist studies. This chapter concludes with a discussion on method and a reflection on the laboratory ethnography conducted in this book.

Notes

1. Cavalli-Sforza *et al.* (1991).
2. Cavalli-Sforza, quoted in Butler (1995: 373).
3. For a critique of this distinction between populations and of the idea that there exist populations that are pure, did not migrate and mix, see Lewontin (1995a: 113). He states: "The notion that there are stable, pure races that only now are in danger of mixing under the influence of modern industrial culture is nonsense." A similar critique is given by Lock (1997, 2001) and Reardon (2001).
4. Roberts (1991).
5. Cavalli-Sforza (1993: 2). This paper can also be retrieved on the Internet at http://www.stanford.edu/group/morrinst/HGDP-FAQ. The author of the Internet copy had become a collective, namely, "The Project's North American Committee." Moreover this copy is a revised version of the copy I received in 1995 from Professor Cavalli-Sforza. Here I refer to the early version of the paper.

6. For example, see de Stefano (1996).
7. Luke Holland (1995). *The Gene Hunters* was a documentary broadcast in June 1995 on Dutch television.
8. NRC (1997); see also Schull (1997).
9. See Lock (2001); Wallace (1998).
10. See, for example, Cavalli-Sforza (1993: 5–6); Dickson (1996). On the debate about patents in relation to the Diversity Project, see de Stefano (1996).
11. Holland (1995). See also Liloqula (1996).
12. RAFI (1993). See also Tutton (1998), who analyzed the fact that the European initiative had received some funding from the European Union whereas the North American initiative was still having problems in terms of their preoccupation with culture and race. Whereas the North American initiative was engaged in a discourse on racism and anti-racism, the discourse of the European initiative was more about culture and cultural heritage in the gene.
13. Reardon (2001).
14. NRC (1997); see also Schull (1997). Schull was chairing the NRC committee and his correspondence was written in response to a news report on the committee's findings, published earlier in *Nature* (Macllain, 1997).
15. As Lorraine Daston (2000: 2) noted, "scientific objects are illusive and hard won."
16. Hayden (1998: 181).
17. This is how the population geneticist Kenneth Kidd recounted his work with Luca Cavalli-Sforza and other colleagues (Roberts, 1991: 1616).
18. After the title of an admirable and a classical book about the ideology of genetics (Lewontin *et al.*, 1984).
19. See for several examples Haraway (1997a); see also Yang *et al.* (1999).
20. See Anon. (2000), "The Genome Special" Issue.
21. In fact, different maps can be made on the basis of DNA: a genetic map and a physical map. Genetic mapping is a technique through which the distance between genes and how they relate to one another can be determined. Physical mapping determines the sequence order of the DNA nucleotides. The goals of the HGP was to determine both types of map for the human genome.
22. Kevles (1992); Lewontin (1993).
23. Gilbert (1992: 96).
24. Kevles (1992: 36).
25. Cavalli-Sforza (1993: 2).
26. Cavalli-Sforza (1993: p. 2–3).
27. See Kevles (1985); Menozzi *et al.* (1994).
28. See Haraway (1991a) for the power of myths and a critique of origin stories.
29. Among these professors in genetics are Mary-Claire King at Berkeley, a former student of Wilson's; Ken Kidd at Yale, who has experience with the growing of cell lines; Ken Weiss at Pennsylvania State University, the chair of the North American Committee. Outside the USA, this list includes, among many others, Julia Bodmer (UK) Walter Bodmer (UK), Alberto Piazza (Italy), Svante Pääbo (Germany), Bertrant-Petit (Spain), Onesmo Ole-MoiYoi (Kenya), Partha Majumder (India) and Takashi Godjobori (Japan).
30. HUGO (1993: 1).
31. At this point, it is important to indicate that there are two conflicting ideas about human origin, with consequences for the reconstruction of human migration history. The most frequently stated theory is the "Out of Africa Theory," the basic hypothesis of which is that all modern humans originated in Africa and colonized the world in one or more migration flows. A second and marginalized theory is the

"Multiple Origin Theory," which assumes that modern humans sprang up in more places and colonized different parts of the world simultaneously. For an example of this debate, see Thorne and Wolpoff (1992); Wilson and Cann (1992). This ongoing controversy is usually reflected in scientific papers, where geneticists tend to underline the fact that their results support the African origin theory. Consequently, now and then, a full paper is dedicated to making that point, such as Wills (1996); Hawks *et al.* (2000).

32. HUGO (1993: 3). See also Hawks *et al.* (2000: 2).
33. HUGO (1993: 28–29). For the purpose of this pilot study, a proposal competition was launched: "Pilot projects for a Human Genome Project: special competition," which can be found on the Internet at http://web.ortge.ufl.edu. This announcement welcomed proposals on "improving techniques for collecting, preserving, amplifying, and selecting DNA markers" and "research on ethical and language issues in a cross-cultural setting."
34. See Cavalli-Sforza (1995: 73).
35. See Tutton (1998).
36. Roberts (1991: 1615).
37. The Professor of Linguistics Colin Renfrew is for this reason very much involved in the Diversity Project. He participated in all the workshops organized by the Diversity Project and was one of the chair-organizers of the Project's most recent workshop: *Human Diversity in Europe and Beyond: Retrospect and Prospect*, 9–13 September 1999, Cambridge, UK.
38. NRC (1997: 13), emphasis added.
39. HUGO (1993: 3); see also Cavalli-Sforza (1991). Defining populations on the basis of linguistic separation is not new. This criterion was, in fact, introduced in the eighteenth century by Johan Fridrich Blumenbach (Hannaford, 1996, especially pp. 202–213; see also Molnar, 1975).
40. NRC (1997: 13); see also HUGO (1993: 12–13).
41. Cavalli-Sforza (1991: 78). Moreover, the correlation between blood groups and languages has already been claimed by C. D. Darlington (1947), a claim that did not sustain criticism; quoted in Molnar (1975: 6).
42. See, Lewontin (1995a).
43. Part of the rhetoric of the Diversity Project concerning the sampling of these populations and the urge to do this as soon as possible is connected to preservationist ideas. These populations are supposedly "vanishing" and "threatened by extinction," in a way losing their value for the purposes of the Diversity Project through admixture. See Hayden (1998) for an elaboration of this rhetoric.
44. HUGO (1993: 7).
45. See also Wallace (1998), especially pp. 60–61, and NRC (1997), Chapter 2. I thank Paul Wouters for bring this point to my attention.
46. Examples of such studies are numerous, but a rather political example presented in a document of the Diversity Project is the following: "One example would be studying Siberian populations to determine whether they manifest any attributes of the susceptibilities of Native Americans to diabetes." (HUGO, 1993: 13).
47. Various examples of such studies are given by Sykes (2001).
48. HUGO (1993: 28); Cavalli-Sforza (1995: 75).
49. HUGO (1993: 17).
50. NRC (1997: 32).
51. HUGO (1993: 3 and 20); see also Cavalli-Sforza (1995: 74).
52. Cavalli-Sforza (1995: 74); HUGO (1993: 22).

53. Polymerase chain reaction (PCR) is the current cloning technology, developed at the end of the 1980s and well established in the early 1990s. About the need for standardization see Cavalli-Sforza (1995: 74); HUGO (1993: 3–4).
54. How technology is implicated in research conducted in the HGP is discussed by Kevles and Hood (1992a); Rabinow (1996a). Especially in the introduction, Rabinow articulates this central role of technology in the HGP and its implications for his own project.
55. HUGO (1993: 3).
56. HUGO (1993: 4).
57. Haraway (1997a: 250), emphasis in original.
58. See also Lock (1997, 2001); Reardon (2001).
59. See also Donna Haraway (1997a: 250). She points out how much easier it proved to slow down or stop the Diversity Project by opposition, compared with the much more powerful HGP.
60. Third World Network (July 1993) http://nativenet.uthscsa.edu/archiv/enl/9307/0036.html.
61. RAFI (1993).
62. For example, the so-called "Blue Mountain Declaration" of February 1995 (http://www.indians.org/welker/genome.html) and the Internet site of the Indigenous Peoples Coalition Against Biopiracy (http://www.niec.net/ipcab); see also Worldwide Forest/Biodiversity Campaign News (http://forest.lic.wisc.edu).
63. For the debate on biological race and racism subsequent to the Second World War, see UNESCO (1951, 1952). For an elaboration on both statements, how they are intertwined with feminist politics, and for situating some of the scientists involved, see Haraway (1992), especially pp. 197–206. On bad science as a post-war category, see, in addition to Haraway, Kevles (1985).
64. See also Star (1995a), who makes a strong point against what she calls the great divide between science and technology, between nature and culture, between social sciences and natural sciences and the like.
65. See, for example, Te Pareake Mead (1996). See other accounts in the same special issue of *Cultural Survival Quarterly* on the Diversity Project.
66. Lock (2001: 80).
67. Eric Wolf, quoted in Lock (2001: 80).
68. Even though genetics since the Second World War, and especially after the UNESCO declarations on race (1951, 1952), has moved away from the concept of race to embraced the concept of population instead, race has never completely disappeared from this field. Race is still a concept of everyday work in medical practice and in genetic laboratories and, in fact, is an issue of debate among scientists (Anon., 1995, 2000; Haraway, 1997a; Annas, 2001; Reardon, 2001; Sanker and Cho, 2002).
69. For a thorough discussion of this protocol and how it helps to stabilize population, see Reardon (2001).
70. Controversies are indeed interesting objects of research because they generate large amounts of material and documents and destabilize scientific facts. Controversies reveal the various networks of individuals, things and ideas involved in scientific facts. Because they become topics of debate, facts that seem hard open up both inside and outside laboratories and show their moulded, fabricated, decided-upon features. Thus controversies disrupt the ordinary, often tedious, get-the-data kind of laboratory work; for controversies in science studies, see Latour (1987); Martin and Richards (1995); Hagendijk (1996). For a variety in approaches towards the Diversity Project, see Joan Fujimura (1998) and Richard

Tutton (1998), focusing on the concept of culture; Corinne Hayden (1998), focusing on kinship and diversity; Donna Haraway (1997a), focusing on purity and contamination; and Jenny Reardon (2001), focusing on the co-construction of a natural and a social order.

71. Needless to say, this notion is not exclusively mine. Various debates in studies of science and technology have been held. For example, there is a whole tradition within gender and anti-racist studies in which the ontological distinction between the object and the subject of research has been questioned. Scientists do not "discover" an object (be this nature or the female body) they have argued but *make* it into one, or "reduce" it to some variables or qualities that can be studied. Additionally, more recently and from a different political angle but with a similar conclusion, scholars who have been studying the process of scientific research have shown that objects do not exist by themselves but are dependent on the very scientific practice in which they are studied. I cite various of this literature in following chapters.

72. On standardization of DNA technologies in order to make them work in different places, see, for example, Fujimura (1992).

73. The notion of performance or enactment was introduced by the sociologist Erwin Goffman (1961) to examine how people stage identities or social roles. The philosopher J. L. Austin (1962) has theorized the performativity of language – "the performativity of utterance" – namely the effect of content and context of words in a specific setting. In the social studies of science, and especially in ethnographic studies, this notion is applied to indicate that objects emerge in networks consisting of people, technologies, language and other things, and to show that objects are "staged" in scientific practice. Moreover, the concept of performance emphasizes the non-stable character of such objects as well as their dependence on locales in which they are actively performed. On the notion of performativity in the social studies of science, see, for example, Mol (2000, 2002); Law (2000, 2002).

2

Technologies of population: making differences and similarities between Turkish and Dutch males

Introducing the argument

This chapter is about population. It will attempt to answer the question: "What *is* population?" Instead of defining it myself or asking geneticists what it is, I want to trace population in genetic practices. I will, therefore, examine how it is enacted in them. To do this, I will analyze a forensic DNA case starting with the basic court case and then discussing the subsequent stages of the case and of the DNA evidence provided. My analysis results in two arguments: first, that geneticists cannot know the individual without a population; second, that in genetics neither the individual nor the population are inherently "biological." Rather, they are *effects* of technologies and routines applied in scientific practice.

As has been discussed in Chapter 1, population is a subject of debate in the Diversity Project. In order to sample, study and compare populations, geneticists aim at achieving a consensus *definition* of what population is. This chapter, however, examines *practices* of population in laboratory routines and reveals different versions of what population is made to be in such locales. In order to know a population, geneticists study cell material from collections of individuals. In forensic science, however, the vantage point is quite different. Forensic geneticists are interested in the individual. Their aim is to identify individual A as similar to or different from individual B. Yet I have chosen this very practice as a site for examining population, for in order to know an individual, forensic geneticists also apply a category of population. In order to produce *differences* (between individuals), geneticists need to presuppose *similarities* (within a population). I will examine practical decisions about individuality and population, and hence about similarities and differences.

Since the main interest of this chapter is in population, little attention is paid to the legal aspects of forensic DNA. Rather than a courtroom, the site of study

A slightly different version of this chapter has appeared elsewhere (M'charek, 2000).

21

is a laboratory, a particular laboratory specializing in forensic science. Since my argument is organized around a forensic case, the narrative will unfold as a journey in which we move back and forth between laboratory and courtroom. I will trace the process of identification and examine the concepts of population embodied in that.

In court

It is 1996 and we are in a courtroom somewhere in the Netherlands. A murder case is in progress. Both the victim and the suspects are Turkish. The victim was kidnapped and killed. Evidence was found in a house next to the victim's body and also in a car belonging to one suspect. The evidence material consists of traces, such as bloodspots, chewing gum and cigarette butts. The evidence at the scene of the crime indicated that more than one person was present. Circumstantial evidence made the prosecutor suspect one individual in particular. The main suspect, however, denied any involvement. Since the victim no longer had a voice, the question remained: can the suspect be identified as the perpetrator?

In court, a relatively new type of expert witness is present to help in the process of identification. The expert witness is a geneticist called in to present and clarify the DNA evidence based on the suspect's cell material and on the evidence: bloodspots, chewing gum and especially the cigarette butts. The DNA evidence consists of a DNA fingerprint. The latter together with a number, 10^{-7}, represents the likelihood that the evidence comes from any other person in the population. According to the expert witness, this number suggests that evidence and suspect DNA coincide. The DNA evidence supports the findings of the prosecutor, which were based on circumstantial evidence. The defence objects to the results of the DNA tests and questions the testimony based on a figure of 10^{-7}.

In order to trace the origin of the results presented in court, what they mean and how they play a role, we can best enter a forensic laboratory. This site is of great importance to the significance of the DNA evidence presented in court. We will first consider this laboratory in the context of DNA evidence in the Netherlands, and then take a closer look at how DNA evidence is produced in that laboratory.

Laboratories involved with DNA evidence

In 1994, a new Dutch law on forensic DNA evidence was passed. This law widened the use of DNA evidence in prosecutions and instituted an infrastructure of sites and regulations concerning the making of such evidence.[1] According to this law, a suspect in a crime carrying a penalty of eight or more years cannot object to DNA testing. In cases where there is at least one other piece of evidence against that particular suspect, DNA testing is compulsory. However, to enforce a DNA test, especially the taking of blood for that purpose, was considered a violation of the person's bodily integrity. Therefore, a number of measures were taken to protect the rights of the suspect. One of these is the suspect's right to apply for DNA counter-expertise. In the Netherlands two laboratories may be involved in the production of DNA evidence. The tests conducted on behalf of the prosecutor are primarily done in the Laboratory of Criminal Justice, Rijswijk (Het Nederlands Forensisch Instituut, Rijswijk), whereas the counter-expertise analyses are conducted in the Forensic Laboratory for DNA Research, Leiden. I will refer to this laboratory as Lab F. If the amount of evidence material does not enable two studies, it is the suspect who may decide which of the two laboratories should conduct the one and only test.[2]

Lab F is part of a broad network concerned with governmental regulations and laws, the Laboratory of Criminal Justice and the Board of Accreditation, the university's Department of Human Genetics and the Sylvius Laboratory in Leiden, pharmaceutical companies and (inter)national networks of forensic scientists and population geneticists, including geneticists in the Diversity Project. In order to reduce the high cost of DNA testing (for prosecutors and suspects) and to make counter-expertise available to all, Lab F is entirely funded by the Dutch government.

The Forensic Laboratory

In this section, we encounter forensic laboratory work. We are introduced to Lab F's rites and rituals, its protocols and procedures, and the particular alignment of technology and trace to produce DNA evidence.[3] Lab F will first be introduced from the perspective of a newcomer, which will reveal materialized institutional arrangements in that particular context and will familiarize us with the laboratory's culture.[4] Then we will learn more about the laboratory's procedures and its organization of work around DNA identification. Finally, we will focus on DNA identification and how this was accomplished in the forensic case introduced above.

Observing the laboratory

I was in Lab F.[5] I was not merely an observer: I had asked for a short introduction to the basics of genetic research. For three and half months, I participated in a project concerned with typing chimpanzee DNA and learned to perform some of the basic tasks of a technician. Since this training constituted my first observational study, I was also learning to observe scientists at work, to study a different culture, to take notes and hold interviews and to develop common ground for an understanding of what was going on.

In order to participate in this laboratory, I was initiated into institutional regulations. Like all laboratory members, I had to sign a medical declaration, I had to be insured against laboratory risks and I had to swear to maintain secrecy about ongoing cases. I was expected to participate in the weekly in-house meetings as well as the weekly joint meetings of Lab F and the Diagnostics Laboratory of Human Genetics. On a more informal level, my daily supervisor and the rest of the members shared with me their accounts of forensics and experiences in the field, which enabled me to enter the discourse of the laboratory.

On my first day, after having been introduced to the laboratory members, the head of Lab F appointed a supervisor for me, explained the project I was going to work on and told me that before the end of the day I would have done my first DNA extraction. In the afternoon, we were indeed extracting DNA from blood. During this laborious work, my supervisor, a technician, told me that he and his colleague were working on two forensic cases and that he was particularly happy that day because he had managed to do a rather difficult extraction. It was nothing like our task, he said, where the identity behind the bloodspots is unambiguous and where the blood is "clean." "His" DNA was extracted from some "dirty" bloodspots on a lampshade. The DNA, so he told me, was still dirty, but he managed to run the PCRs (DNA copying technology) necessary for identification.[6]

I learned that the complexity of extractions and the social relevance of identification caused all technicians, without exception, to prefer working on forensic cases rather than on the research projects of the laboratory. Whereas most of the technology applied by the technicians was standardized, the starting point of the cases, the extraction of DNA, demands insight, experience and care. At the beginning of the procedure, the technicians have to assess the evidence material and gain an idea of how many extractions can be made from it, or whether it is possible to extract *any* DNA *at all*.

It is also at this stage that the cases acquire colloquial names. Since the laboratory members do not know the legal details of the case (all cases are assigned a registration number once they enter the laboratory), they have developed a laboratory-specific typology. The cases could be referred to in terms of the registration numbers. However, in practice, they are attributed more communicable names, such as the case of the lamp shade, the case of the stamps, the case of the bracelet, the rape case, the blackmail case, the paternity case, and even a case called number 9 gains another meaning in this context. These attributions may contain ethnographic information.[7] Our case had acquired the adjective "Turkish." The ethnographic contents of this adjective will be addressed later in the chapter.

The Turkish case was closed before I came to Lab F. Yet it continued to be mentioned on various occasions and it became clear to me that the case was important to this laboratory. The material for this chapter is based on many conversations during my training, some of the case material that could be made available to me and interviews I conducted towards the end of the training.

Laboratory practice

Lab F is a predefined environment in terms of protocols, technology, knowledge and space. Any step in the analysis is recorded on specified forms on which the case number, the name of the technician as well as information about the utilized chemicals (such as "lot numbers" and expiry date), kits and technical devices appear. Also all analyses and technology applied are carefully defined in the various protocols. These measures are aimed at the transparency and repeatability of any DNA test, even after years have passed. One of the major concerns in this respect is the confusion or contamination of samples. Prompted by concerns about contamination, the most pivotal spatial division is the division of the laboratory into two areas, referred to as "pre-lab" and "post-lab."

Contamination is one of the major worries of all forensic laboratories. In 1996, the US Committee on DNA Forensic Science produced *The Evaluation of Forensic DNA Evidence* on behalf of the NRC in Washington, DC.[8] This report, which aimed at redirecting forensic DNA research, addressed the risk of contamination at the different stages of the identification. "*Contamination* has been used as an umbrella term to cover any situation in which a foreign material is mixed with an evidence sample. Different kinds of contamination have different consequences for the analysis."[9] The committee recommended a number of measures to reduce the risk of contamination in forensic work.

In the laboratory, I learned quite soon that measures against contamination were taken very seriously. On the Wednesday morning of my second week, I was "setting up a PCR" in the pre-lab. I had a question about the storage of reagents and, since my supervisor was working in the post-lab, I went over there to find him. In the doorway, however, I froze on hearing a chorus of voices shouting: "Laboratory coat! Laboratory coat! Take it off!" I looked down and realised I had forgotten to take off my pre-lab coat while planning to enter the post-lab. Of course I knew rule number one: equipment for the pre-lab should remain there, and if it enters the post-lab it cannot be taken back without extra effort. The movement from pre- to post-lab is easy but the other way around requires extra measures (sterilization of instruments, putting on of gloves or a laboratory coat). My overt confusion made the laboratory members laugh and, after a moment of despair, I closed the door and ran back to take off my coat in order to keep it free from (post-lab) contamination. From that moment on, the risk of contamination became very real. The certificate of the Board of Accreditation hanging on the wall became more meaningful and serious. I noticed the friendly but critical eye of the quality control manager much more often.[10] Furthermore my laboratory coat, the rubber gloves, and the mask I occasionally wore became strict borders between foreign material and the DNA on which I was working. What is foreign does not have to be strange. My supervisor told me that all laboratory members have their DNA profiles typed.[11] This information enabled the staff to trace the source of potential contamination and to exclude the possibility that the occasional foreign material is theirs.

The pre-lab is the more sensitive environment. This is the space where the cell material of all cases comes in and where DNA is extracted. The extracted DNA sample remains in this laboratory and is used in small quantities for the different analyses. For each of the analyses, the DNA will have to be copied using the PCR copying technology. With the help of a thermostable enzyme (polymerase), the PCR machine produces a million-fold copy of a particular DNA fragment, and so enables its visualization. This copying procedure, also called DNA amplification, constitutes the very division between pre-lab and post-lab: the names refer to pre-amplification and post-amplification. Thus, the PCR machine sets the boundary and is placed in the post-lab, where the amplification takes place. Before the DNA leaves the pre-lab for the purpose of copying, it is mixed with additives necessary for that step, such as nucleotides (DNA building blocks), primers (synthesized DNA fragments), the enzyme and other PCR chemicals. This mix of chemicals and the PCR amplification is pow-erful, making the copying procedure sensitive to contamination by free-floating DNA fragments, which are more likely to be found in the post-lab. Therefore, the mix is prepared in the pre-lab in a "flow-cabin," where such floating DNA fragments are least likely to be found. Before the mixture leaves the flow-cabin, it is placed in lidded cups. As a routine check of possible contamination, with each PCR reaction a positive control (a DNA sample for which the information is already known) and a negative control (usually double distillate water) join in every step when typing the DNA of a case.

Once the technician enters the post-laboratory with a rack of cups con-taining the mixtures, the chemicals have run their primary course. The sub-sequent experiments are conducted carefully in order to avoid mistakes when adding other chemicals to post-PCR DNA solution (the so-called PCR prod-uct) and to avoid interchange between the cups. The results, consequently, are dependent on the right tools being used and on the results being interpreted correctly.

The emphasis on reducing the possibility of contamination is essential to most laboratories in the field of human genetics, but it is perhaps even stronger in the context of Lab F. This laboratory is a *forensic* laboratory: the results of experiments conducted here have decisive consequences for the imprisonment or liberty of a suspect and for the kinship or identity of an individual. Mistakes or ambiguities in the DNA analysis are reviewed and should not appear in the final report.[12] Lab F operates predominantly on behalf of the suspect or the accused (the alleged father, blackmailer, murderer or rapist). In analyzing DNA, this laboratory studies the same tissue, hair or blood as that studied on behalf of the prosecutor in the Laboratory of Criminal Justice. Reports on both studies are then submitted and presented in court. As stated before, if there is not enough evidence material to support two studies, the suspect may choose where the only possible DNA evidence should be produced.

Now we will follow the procedure of DNA identification. We will learn more about what counts as evidence and what does not, and about the possibility of identifying a unique individual. Our focus will be on the case of the Turkish suspect and, for reasons that will become clear later, at this stage it will be referred to as the *T-case*.

The T-case: DNA profile typing

In the T-case, there were two suspects. Although one was more suspect than the other, the cell material of both was supplied for DNA analysis. In this case, the amount of evidence cell material did not support two studies and the defence requested Lab F to produce the one and only DNA profile.

According to protocol, the evidence material did not travel alone. It was accompanied by a short description of the case and of the material. In order to guarantee the privacy of the suspect, only the head of the Lab F receives this information. The technicians are not supposed to know the names of suspects, nor where the crime in question took place. The latter is an extra measure to prevent the technicians from forming a biased view in cases where a crime becomes a public issue.

The T-case was treated routinely. Two technicians conducted two parallel and independent analyses, from the extraction of DNA from cell material to the typing of the DNA profiles. DNA profiles of the evidence material and of the suspects were typed in order to check for a match between them. The profiles are compounds of *genetic marker* information. A genetic marker can be seen as a small fragment of DNA that may vary in length between individuals, according to the number of nucleotides (i.e. the number of DNA building blocks) it contains.[13] The various lengths that can be found in different individuals are referred to as *alleles* (allelomorphs). For example, an individual A who carries a sequence fragment of 290 base-pairs may be said to carry a different allele from that of an individual B, whose sequence fragment is 294 base-pairs.[14] It is this variation in sequence length, or alleles, between individuals that makes genetic markers useful for profile typing and for comparing individuals with one another.

In order to produce the DNA profiles, Lab F typed three groups of markers: five *poly-markers*, one *HLA* (human leukocyte antigen) marker, and one *short tandem repeat* (STR) marker. The poly-markers are a standardized package of five markers located on five different chromosomes; they show polymorphisms (i.e. variations between individuals) based on one single base-pair substitution in each DNA fragment.[15] HLA markers are located in several hundred genetic sites on chromosome 6 and are responsible for the antibody system. The STRs consist of short sequences of two to five nucleotides that repeat in tandem, such as the tandem CTAT, which may repeat eight to twelve times, or ATT, repeating 10 to 16 times in different individuals. This set of markers, poly-markers, HLA and STRs, is informative for the very reason that their appearance may differ between individuals, either in length or in sequence composition. However, there is a fair chance that two individuals might "look alike" for one of the markers; using more markers reduces this probability and produces a more individual profile. By typing the whole set of markers, the laboratory obtained individual profiles of suspects and evidence DNA.

Based on the DNA analyses, the profile of one of the suspects matched the evidence DNA. However, a match by itself does not imply identification. It does not automatically link the biological trace to the suspect and it does not contribute to the identification of the suspect as the perpetrator. More work is required to achieve that. In order to be sure that the match between the profile of the suspect and that of the evidence DNA was the most likely, a forensic laboratory would compute a matching probability based on profiles in the control population (a *matching likelihood number*). This step is crucial. Forensic work is based on the presupposition that the suspect is innocent and that the evidence DNA was left by somebody else at the scene of the crime.[16] Therefore a laboratory has to estimate the chance of a match between evidence DNA and any other individual in the population. Since it is not feasible to type the profiles of all the individuals in a population, forensic laboratories work with a control population based on a "random" collection of people. For Lab F, this was a collection of Dutch people. In calculating the matching likelihood, Lab F compared the profile of the evidence DNA with those of 168 males and females that were available in its databank. On this basis, a probability was calculated expressing the chance that the marker profile of the evidence DNA could be found in the population at large. The procedure for this is as follows.

Suppose that for three markers the specific fragments found in the evidence DNA, the *alleles*, could be found in the databank in the following percentages: allele for marker I is represented in the databank by 10%, allele for marker II is represented by 5% and that of marker III can be found in 2%. To calculate the chance that the three marker fragments (alleles) combined may occur in the population, the frequencies are multiplied. In this specific example, the chance would be: $10/100 \times 5/100 \times 2/100 = 100/10^6$. Hence the chance of a match is 1 in 10 000. This number is called the matching likelihood number and is, therefore, the result of a simulated comparison between the suspect's DNA profile and that of the population at large: that is, the population of which this individual is considered a member.

In the T-case based on seven markers and Lab F's control population, a matching likelihood of 10^{-7} was calculated. In other words, the chance was 1 in 10 million that the profile of the evidence DNA would match that of any other individual in the population from which the samples were drawn. This calculated probability makes the DNA profile of the suspect into a *DNA fingerprint*. The figure 10^{-7} was considered acceptable according to the standards of Lab F and to those of the court.[17] The figure was, therefore, presented in court as evidence of an exclusive match between suspect and evidence DNA, and thus as the identification of the evidence material. One could say that 10^{-7} was the basis for the DNA evidence in this case. A higher matching likelihood would counter the testimony of the DNA comparison and challenge the authority of the DNA fingerprint.

For Lab F, this chain of procedures usually ends when the information is sent to court. The DNA profiles, the matching likelihood number and the methods of analysis are recorded in a report drawn up by the secretary and, after a simultaneous check of the report by a technician and the head of the laboratory, it is signed, sealed and sent to the court. A copy of the report printed on red paper, signaling that the case is closed, remains in the laboratory. In our case, the head of Lab F was invited to the hearing as an expert witness to present the results and to answer any questions.

From this discussion it becomes clear that DNA fingerprints are technical products, intertwining a specific nature of individuality and of population. We will now go back to our case and follow the technicality of both categories. As indicated before, the defence had problems with the results presented in court. Let us now have a closer view at these objections and at how these relate to the T-case.

Back in court

In court, the DNA tests of Lab F supported the findings of the prosecutor. These tests established a link between the evidence, found in the car and in the house next to the victim's body, and the main suspect. However, there appeared to be a problem. The defence did not accept the DNA evidence and started to raise questions about how the matching likelihood number was produced, how the comparison was done and about the control population upon which the figure 10^{-7} was based. Given that information, the defence argued that their client was not just a suspect, but a Turkish suspect. Their client was, after all, of Turkish descent. Did Lab F take that into account and could it guarantee that the control population was representative of their client's DNA profile? What do these objections mean?

In order to calculate the matching likelihood number and to produce the DNA fingerprint, Lab F had compared the profile of the Turkish suspect with those of a Dutch control population. As stated before, the comparison was based on the presupposition that the suspect was not guilty and that the perpetrator was "out there" in the population. Therefore, theoretically, any other person in this population has an equal chance of having committed the crime in question. The defence questioned whether this presupposition was still maintained, given the "Turkishness" of the case. It argued that since the victim, the suspect and other individuals at the scene of the crime were all apparently of Turkish descent, it would not be plausible to presume an equal probability that any individual in Dutch society could have been the perpetrator. Questioning the control population used means doubting the matching likelihood number of 10^{-7} and, consequently, questioning the validity of the DNA evidence. Whereas the defence emphasized the Turkishness of this case, Lab F had not been informed beforehand about the descent of the individuals involved. While their names might have suggested it, no special attention was paid to the possibility of non-Dutch descent. Lab F used its Dutch control population *as usual*. Thereupon, the defence raised the question of whether one could presuppose the absence of differences between Dutch and Turkish genetic makeup.

The scepticism of the defence raised a number of problems for the court. Would the matching likelihood be equally low if the evidence DNA had been compared with a Turkish control population? Did Lab F use the appropriate control population to determine the DNA fingerprint of the suspect? Consequently, could we be sure that the match between evidence and suspect DNA (*inclusion*) meant that the suspect is indeed the perpetrator (*identification*)?[18]

Although the traces of DNA based on evidence in the car and at the scene of the crime seemed to match with the profile of the suspect, the court decided that the matching likelihood probability contained no evidential power without further information about a more "appropriate" control population. Lab F was asked to take the Turkishness of the case into account, to investigate the possibility of using genetic data of a Turkish population and to recalculate the matching likelihood number for establishing the *identification* of the suspect.

We shall see below how Lab F responded to this request. Furthermore, it will become clear that answering these questions opened up a broad field of assumptions, procedures and negotiations crucial to forensic work: to population genetics as well as to forensic DNA evidence. To understand what is at stake in our case, let us make a brief detour to another forensic case where the DNA evidence was challenged.

Expert and counter-expert

Unlike the USA and the UK, in the Netherlands forensic DNA did not become an issue of public debate until the mid 1990s. Before introducing the 1994 amendment to the Dutch Criminal Code permitting DNA evidence in lawsuits, a committee was appointed to investigate its impact and assess its future use. In that very context, discussing the reliability of technology, references were made to controversies in the USA and the UK. The committee considered technology to be no longer unreliable and spent most of its time regulating the rights of the suspect.

However, despite the absence of controversies in the Netherlands, the defence may have been informed about controversies surrounding DNA evidence abroad and may have intended to use these as jurisprudence.[19] One such controversial case took place in 1990 in Franklin County in Vermont, near the Canadian border. This was a rape case in a caravan camp of Abenaki Indians.[20] The judge in this case discarded the DNA evidence because it was obvious that not all inhabitants of Franklin County had an equal chance to access the caravan camp or, in other words, there was a higher probability that the perpetrator was a member of the Abenaki Indian community. Franklin County is ethnically very mixed and the Abenaki Indians make up the largest population. Although there were genetic databanks of other populations in the county, there was none for the Abenaki Indians. Therefore, the suggestion to compare the DNA profile of the suspect, who was an Abenaki Indian, with various other ethnic groups living in the region was deemed inadequate and unreasonable. The DNA evidence was deemed inadmissible and the court dismissed the case.

Emphasizing the need for an appropriate control population, the defence in T-case seemed to call forth a similar dilemma. Before returning to the case, however, we will take a closer look, first at matching likelihood estimation and DNA fingerprinting, and then at the different concepts of population that have been presupposed in the case thus far. With reference to matching likelihood and DNA fingerprinting, we will try to understand the traffic of DNA evidence between laboratory and court by applying a theoretical notion, that of an "immutable mobile."

Matching likelihood numbers and DNA fingerprints: immutable mobile?

Bruno Latour raised the question: ". . . how can distant or foreign places and times be gathered in one place in a *form* that allows all the places and times to be presented at once, and which allows orders to move *back* to where they came from?"[21] in his *Drawing Things Together* where he addressed inscriptions of scientific practices and "worlds" onto objects. The objects he considered can be seen as representation conglomerates in which many worlds and practices are reassembled in such ways as to make these presentable in a different setting. Such objects are transportable, and unchangeable during their transportation. What "form," then, should these objects take to do this job? Latour suggested that we think of graphs, models, figures, written texts, specimens or samples, which combine various practices in a specific material "form." In this fixed condition, they are immutable, and they are mobile because they are either easier to transport or to reproduce. They are thus "immutable mobiles." Immutable mobiles, as representational devices, allow scientific practices to be transported from one laboratory to another and from one field to another. Latour argued that immutable mobiles are "things" gathered, displaced, made "presentable" and convincing to those who have not been there. These "things" may combine and recombine different fields, because they are made flat and reproducible.[22] Here I will consider matching likelihood numbers and DNA fingerprints as immutable mobiles.

In forensics, a DNA fingerprint cannot exist without a convincing matching likelihood number. Without this number, the fingerprint would be just another DNA profile. Although the two accompany each other on a journey from laboratory to court, matching likelihood and DNA fingerprint reveal a different reality about where they come from.

Classifying a DNA profile as a fingerprint suggests an analogy with conventional fingerprints. This analogy does not always hold. The print of a finger of a *Drosophila* or a dog does not exist, yet both may have a DNA fingerprint. But the fingerprint analogy is instructive of how to understand the DNA evidence.[23] The understanding implied is that DNA fingerprints are easy products. In addition, the suggestion is that even in daily life anybody can produce his or her own fingerprint, whether based on DNA or the print of a finger. Paul Rabinow described the convenience of conventional fingerprinting as follows: "Little skill was necessary to obtain fingerprints and not much more was demanded to classify them. No confession was required, a physical impression would do."[24] Thus, understanding a DNA fingerprint in terms of a conventional fingerprint mobilizes a previous and successful individuality-determining practice in forensics. While facilitating this reading, the analogy seems to prohibit another, namely a complex view of DNA fingerprinting. What knits the two practices together also seems to set them apart: this is the belief that ever-better technology and knowledge are produced to do the same job. It is the belief that DNA typing has superseded conventional fingerprinting and become "the ultimate identification scheme."[25] This feature of DNA fingerprints is emphasized by the matching likelihood number.

The second immutable mobile, the matching likelihood number (especially an impressive number like 10^{-7}), mobilizes a scientific practice. It mobilizes a complexity, hidden from the view of those "who haven't been there." This scientific practice is a number-producing machinery that produces facts out of human tissue. A stranger does not (need to) know what is going on in this machinery, how DNA is acquired, "unravelled" and made into a DNA fingerprint. The matching likelihood number mobilizes and establishes the complexity of these procedures as well as the scientific prestige attached to the production of DNA evidence. The information that comes out of the laboratory is a DNA fingerprint with a number.

The DNA fingerprint, as an immutable mobile, transports into the courtroom, a (scientific) world where fingerprinting is accepted as an individuality-determining practice. The matching likelihood number, by comparison, mobilizes a scientific practice where numbers as facts are produced out of human cell material. Whereas, the former suggests familiarity, the latter evokes strangeness.[26] If we put this in the context of our case, it is exactly this strangeness the defence sought to challenge. The objection of the defence can be seen as an objection to both the matching likelihood number and the DNA fingerprint presented. Thus, the defence questioned the immutability of these mobiles and opened up the practices inscribed onto them and transported through them.

How to read the individual profile or the DNA fingerprint is now the central issue and has become a matter of comparison between an individual and a population. Both sides of the comparison are of great importance. As already indicated, the main focus of this chapter is population. So let us try to trace the different concepts implied in this part of the case. Special attention will be paid to where differences and similarities are located and how similarities and differences between Dutch and Turks contribute to different concepts of population.

Similarities presupposed

The first concept of population is indicated by the control population of the laboratory. The control population of Lab F was based on three population samples. Combined, they were deemed to be representative of the Dutch population at large. The sampling procedure of the control population contained specific notions of what counts as population.

Of the 168 samples in use in Lab F, 50 males were selected from Dutch hereditary-diseased families at Department of Human Genetics in Leiden. The samples came from healthy men related by marriage to the diseased families. These samples were selected by family name only.[27] A second set of 50 female samples was randomly drawn from a large study on contraceptives in Dutch women; while the samples were drawn at random, the collection as a whole was known from elaborate medical records. These samples, as well as the third set of 68 male samples, were made available by the TNO (Dutch Organization for Applied Scientific Research).[28] The 68 male samples were also drawn randomly from a larger number of samples: 2018 35-year-old males taking part in a large-scale TNO study of (susceptibility for) heart disease in the Dutch population. This study was conducted in three regions of the Netherlands, represented by the towns of Doetinchem, Leiden and Amsterdam.

Hence the control population of Lab F as a whole was based on both genealogical ties (as expressed in family names) and on specific ties to the Dutch medical system.[29] Although the medical records contained a large amount of information about the individual samples and about the sampling procedures, the medical ties are not of particular importance to Lab F. In compiling the control population, the laboratory was not so much interested in those individuals as such but rather as representatives of a population as a whole. Yet the choice for healthy men, in the first collection, is indicative of a practical consideration: the laboratory wanted to compile a *normal* control population. Not a population of individuals living in the Netherlands, but a *normal* Dutch control population was the goal. Names, just like health, were treated as devices to produce homogeneity in that population. Furthermore, since the control population consists of 168 samples only, it was of statistical significance to know that a genetic profile was not doubly represented because two individuals belonged to the same family; the way to reduce this chance was by excluding samples that came from individuals sharing family names.[30]

What, in view of the above, were the practices embodied in the sampled population, the source of the control population for Lab F? On the one hand, the choice of names as an overall criterion for samples guaranteed the representativity of the samples. Individuals who shared the same name were excluded. On the other hand, it also suggested that family names make population. A population on this basis consisted of individuals linked by names. It indicated that Dutchness is in the name. Names function as an attempt to capture an unambiguous genealogy of what Dutchness is, and consequently of what a population is.

A second concept of population can be traced in the objection of the defence, and the emphasis placed on the Turkishness of the case. At stake is the matching likelihood number and the basic statistical presupposition it expresses. The number presented was the result of a comparison with a Dutch control population, and presupposed that any citizen of Dutch society had an equal chance of having been at the scene of the crime. The defence opposed this by questioning whether this presupposition would still hold if suspect and victim were Turkish. As indicated above, Lab F had not been informed about the descent of these individuals: the information contained their names only. Their names would, of course, suggest a different descent, but this was not taken into account. Did the laboratory have a particular reason not to do so? However, in accord with forensic science, Lab F also works with the presupposition that the suspect is innocent. Comparing the evidence DNA profile with that of the control population means looking for a possible match, other than the match between the suspect and the evidence DNA. Despite the fact that more non-Dutch individuals were at the scene of the crime – which reduced the chance that any citizen might theoretically have been there – the laboratory considered its control population to be representative. Does this mean that the laboratory was ignoring the statistical proposition of forensics: the possibility that any other person could have committed the crime?

Seen from a different angle, taking the conduct of Lab F seriously could lead to a different conclusion. Namely that the laboratory had taken this proposition into account but had failed to make a distinction between what might count as Dutch and what might count as non-Dutch. From this perspective, the practices that enabled the production of the laboratory's databank (the samples drawn from various groups of individuals and decisions for particular samples and not others) move to the background and the data come to represent a much larger population. Even more, the data had come to represent not only a Dutch population but a more general population. What concept of population would include Dutch and non-Dutch as belonging to one category? Here, the laboratory practice provided some clues. For Lab F, the control population was the databank *as usual*, and the suspect was a suspect *as usual*. A T-suspect is not a common occurrence in this laboratory.[31] Having compiled its databank with care, the laboratory had grown to view it as the control population full stop. The databank had thus become "a black box." The stake in this particular black-boxing is not the content or the composition of the databank but the daily routine of forensic work in which it is used.[32] Normally, the databank does the task of a control population quite well and is, therefore, "reflexive" of routine practice.[33] One could say that, through daily routine, the laboratory seemed to have developed a blind spot for T-cases. Therefore, the second concept of population is an effect of routines and everyday practices. In such practices, the population had become the control population *as usual*.

Proposing differences

For the benefit of its client, the defence questioned the matching likelihood probability based on the control population of the laboratory. This indicates a *third concept of population*. Whereas Lab F presupposed similarities between Dutch and non-Dutch, the defence put forward the possibility of differences. By pleading that the case should be viewed as a Turkish case, the defence questioned the "representativity" of the control population applied. Could the identification of the Turkish suspect be established on the basis of the proposed matching likelihood number, or would the profile be more common if compared with a Turkish population? Challenging the claim of *identification* embodied in the matching likelihood number calls upon the issue of genetic distance. Are Turks and Dutch genetically close enough to be situated in one population, or does a difference in descent indicate a genetic distance between groups of people?

These objections can be viewed within the realm of population genetics, where it is argued that groups of people from different parts of the world may differ in genetic makeup. Such differences often concern the frequency with which alleles occur within specific populations. In this respect, a matching like-lihood number that is *exclusive* within the context of a Dutch control population might be less convincing if comparison were based on a different population. Emphasizing the fact that the suspect is of Turkish descent, the defence pleaded that these differences should be taken into account.

This objection suggests a concept of population based on genetic proxim-ity and distance. Individuals are members of a population when their genetic makeup/profile is represented in this population. Thus people from different parts of the world are included in or excluded from a population on the basis of genetic nearness or distance.[34]

From these analyses, it became clear that in forensic practice there is no such thing as *the* population, that different practices of population may be in play at the same time, and the specific concept produced has consequences for individuality. To know the individuality of the Turkish suspect is thus dependent on population. We shall now go back to Lab F and view how it responded to the court's request: to solve the problem of the control population and to seek comparisons with a Turkish population.

Back to the laboratory

By now the laboratory's control population had become a Dutch control population, while the T-suspect had become a Turkish suspect. It had also become clear that it was not possible to identify the Turkish suspect on the basis of routine and standardized laboratory practice. Matching likelihood, control population and DNA fingerprint had all become matters of more decisions than usual. Studies of Turkish populations were, therefore, gathered and considered. Two published papers were found to be of specific importance; I will refer to them as the "German study."[35] Lab F tried to meet the request of the court by calculating a new matching likelihood number based on the Turkish data referred to in the German study. In an interview I held with him retrospectively, the head of Lab F stated the following about the questions of the defence.

> The question of "representativity" raised by the defence is a relevant question. In Germany, a study was conducted measuring the allelic frequencies of two different Turkish groups: Turkish migrants living in Brussels and Turkish groups living in the Adana region (south-eastern Turkey). The suspect is also from the south-east of Turkey. They are all Caucasians by the way, just like the Dutch. For most genetic features there is little difference in the allelic frequencies among Turkish people, but if compared with Dutch males, differences do occur.[36]

As a response to the defence, Lab F compared the DNA profiles of the suspect with the data of the German study. With the information about the allelic frequencies of the Turkish population contained in that study, a new statistical analysis was carried out for the suspect DNA. Hereupon, the matching likelihood was recalculated as 10^{-6} instead of 10^{-7}. This means that the chance that the profile of the suspect would match with any other Turkish person has grown to one in a million instead of one in ten million.

Back in court

In court, Lab F presented the new results, which showed that comparing the allelic frequencies of a Turkish population with the alleles of evidence and suspect DNA allowed it to calculate a new matching likelihood probability. The papers that provided the laboratory with the data were included as scientific evidence. Again, the defence was not convinced and stated that even the matching likelihood number based on the data of the German study may not have accurately represented the case at issue.

The objection of the defence at this point was not so much based on the representativity of populations, as on that of genetic markers. Whereas the markers used in the German study contributed to the conclusion that there is no difference among Turks, the defence aimed at just the opposite. It had mobilized more data about the Turkish population living in Brussels and argued that Turkish populations may be similar for some markers but can differ for others.[37] Since the markers used in Lab F and those in the German studies differed considerably, the defence still had doubts about the matching likelihood presented. The German study, for example, was conducted solely on the basis of STRs, whereas the set of seven markers used in Lab F included only one STR marker. Furthermore, the six other markers used are found in so-called "coding regions" of the DNA, which also makes them sensitive for population substructures. The defence suggested comparisons with the markers and data used in the Laboratory for Criminal Justice, Rijswijk.

Tools of similarities and differences: genetic markers in DNA fingerprinting

At this stage, genetic markers became an issue of debate in DNA evidence. Genetic markers are key categories in population genetics.[38] In forensics, genetic markers are also subjects of great debate and discord, as they are often matters of life and death in such cases.[39] Likewise, in our case, markers were important actors that kept popping up. So let us examine them more closely in order to understand the stakes involved in them for this case.[40]

In studies of genetic diversity as well as forensics, genetic markers are selected on the basis of three criteria. First, markers in the non-coding region of the DNA are preferred over those found in the coding region. Contrary to non-coding DNA, coding DNA has crucial functions in the cell because it helps to produce proteins. The pattern of change in coding DNA occurs through processes such as the accumulation of mutations, which over a period of time is the source of genetic diversity, and the way such mutations are passed on to offspring may be restricted. This is because minor changes in coding DNA may have devastating effects on the proper functioning of the cell and, therefore, on the health of an individual.[41] By contrast, in forensics, the underlying and most important supposition concerning markers is that such changes are not restricted and that they are inherited at random. This implies that genetic diversity and the various different alleles of a marker are randomly distributed within a population. This is based on the so-called assumption of "random mating," according to which people choose their partners at random and also pass on their genetic material (and the diversity that can be found in it) at random. In *The Evaluation of Forensic DNA Evidence*, it is argued that "for some traits the population is not in random-mating proportions. Mates are often chosen for physical and behavioural characteristics . . . For example, people often choose mates with similar height, but unless a forensic marker is closely linked to a possible major gene for heights, the forensic genotypes will still be in random-mating proportions."[42] Therefore, the preference for forensic analysis is for markers in non-coding DNA and which are not linked to genes. The so-called "discrete alleles" of such markers are supposed to meet the condition of random mating and, therefore, fit into the statistical models that support this proposition. This does not mean, however, that markers linked to coding DNA are necessarily excluded. The HLA and some of the poly-markers used in our case, for instance, are linked to functional DNA. Considering the random mating issue and the choice of markers, geneticists strive for markers that are not linked to each other and that are inherited independently. Preferably, they look for markers on different chromosomes, as in the case of the five poly-markers.[43]

Second, a marker should be polymorphic within a given "population." This means that it should show different alleles within the sampled population. The discriminating power of a marker depends on the number of its alleles. If the number is too low, this will enlarge the chance of a match between two genetic profiles. Therefore, using markers with a low number of alleles requires the use of a larger databank (a few hundred or more).[44]

Finally, a marker should not be *too* polymorphic. This has to do both with other types of case that one can find in forensic laboratories and with the statistical models used. If a marker has a high number of alleles, it would be an interesting marker for identification: the more variation the smaller the chance of a match between two individuals. However, forensic DNA is not only concerned with identifying possible criminals but also with paternity testing. For the latter type of profiling, markers with a high number of alleles are especially problematic. Many alleles or a high variability indicate that a particular DNA fragment changes or mutates relatively fast. Mutations may even occur between two generations, distorting the results of paternity tests, which are based on similarities and differences between parents and offspring. For practical reasons (such as being able to use the same set of marker for both DNA evidence and paternity testing, not having to train laboratory members to work with too many different markers and for economical reasons), markers are chosen that are polymorphic but not *hyper*variable. This choice is also involved in statistical models. For example the models of Lab F do not take into account the occurrence of mutations between one generation. However, if Lab F chose hypervariable markers for the purpose of identification, it would have to change its statistical models to be able to use them. Therefore, the practicability of markers in a specific laboratory context puts constraints on which of these will be considered *good* in forensic DNA practice.

Now that we have outlined some of the criteria of good genetic markers in forensic research, let us go back to the T-case and have a further look at the different concepts of population that have been touched upon in the previous section.

Arguing for similarities

A fourth concept of population can be traced in the interview excerpt quoted above. When talking about the problems of representativity, the head of the laboratory mentioned in passing that Turks, like the Dutch, are Caucasian.[45] This notion of population seems to warrant the first and prior approach of Lab F not to distinguish between the two; where population was their "local" population as usual (the second concept of population). However, as I have indicated, the laboratory was not informed about the descent of the individuals involved. Taking their names for granted suggests that the control population, which was based on Dutch names, became a self-evident part of a routine practice and using Caucasian in this context should be understood in other terms. Caucasian is here a racial category, suggesting a taxonomy of population based on race.[46] Race is an ambiguous but nevertheless relevant category for geneticists.[47] The Caucasian, Negroid and Mongoloid races are seen as the three main races of the world.[48] According to this taxonomy, "races" within the three main races are called "population substructures" or "subpopulations."[49] In a way, this suggests that "population" is nothing but another term for race. Even though genetic technologies blur clear-cut categorization along racial boundaries (whatever these may be), races are entry points for genetic studies (sampling procedures and comparisons) as they are embedded in a long history of research in this field.[50] In this concept of population, Turkish and Dutch are included in one race, namely Caucasian. It suggests a practice of population in which racial boundaries coincide with population boundaries, a classical biological basis for similarities and differences.[51]

A fifth concept of population draws upon the German study, as I have referred to in this case. On the basis of the German study, Lab F could draw the conclusion that Turks are not Dutch when it comes to their genetic material. The German study showed that allelic frequencies differ more between Germans and Turks than they do among Turkish people. Two Turkish groups were studied, one of which lived outside Turkey. This group had migrated to Brussels in the 1960s and had been living there ever since. No indication is given of the group's place of origin in Turkey. That did not seem to be an issue. The conclusion in one of the papers is that "neither Turkish *subpopulation* showed any significant differences for any of the three STRs, indicating that the time of geographical separation was too short to have had an influence on the allele frequencies."[52] Since the study answers the question of whether migration had had an impact on genetic "homogeneity," it becomes clear that "homogeneity" is located within a national context. Calling the two groups *subpopulations* indicates that they are derived from one *population*, an overall Turkish population. Hence what is Turkish is correlated with being a subject of the nation state Turkey. The national boundaries of Turkey, therefore, contributed to what may be seen as Turkish. This concept of population suggests that it is a matter of national boundaries.

More generally, within the realm of population genetics, national boundaries are seen as prohibiting conditions for "random mating" between members of different nations/populations and as enhancing random mating within the nation/population. The spread of alleles is expected to be higher within national boundaries.[53] The scepticism of the defence could be rephrased as addressing exactly the presumption of an "easy" spread of alleles within national boundaries.

Arguing for differences

The defence's objection to the presumed distribution of alleles within Turkey introduces *the sixth concept of population*. Contrary to the presumption of similar allelic frequencies within state boundaries, the defence presented data that showed just the contrary. Therefore, the idea that the country as a whole may have a general allele frequency representative of Turkish individuals at large is open to question. In addition, since the markers used in Lab F differed considerably from those studied in the German paper, the defence asked for more comparisons. As a check, it suggested a comparison with the markers and data used in the Laboratory for Criminal Justice, Rijswijk.

To doubt the content of 10^{-6} from this perspective suggests that populations may be tied to specific markers. Depending on which markers are used, population may be produced differently. Depending on markers, alleles may be equally spread over the whole world, they may be clustered in specific patterns, or they may be found in one population and not in another. The matching likelihood number correlates to the frequency of alleles in a given population. This means that to be able to say anything about matching probabilities one needs to be sure that the specific profile is represented, in terms of alleles, in the population. For, if the suspect is a carrier of an allele B and allele B happens to be common in the appropriate control population but unusual in the one used, then the matching likelihood number will be biased and will tend to show a lower figure. On this basis, the defence demanded a clearer answer about the clustering of alleles and how genetic makeup is affected when different markers and different data about the control population are being applied.[54] From this, it can be stated that population is a product of genetic markers. Depending on what types of marker are used, populations can be clustered anew.

Before going back to the laboratory to see how it answered the question of genetic clustering in populations, let us take a second look at the matching likelihood number and the DNA fingerprint as *immutable mobiles*. We will focus on the stakes in their immutability and the effect of their mobility in this particular case.

Matching likelihood numbers and DNA fingerprints: immutable mobiles

Earlier in this chapter, I suggested that DNA fingerprints and matching likelihood numbers can be viewed as immutable mobiles. Numbers are immutable mobiles par excellence and, as I suggested, analogies bear this power as well. Both have the capacity to mobilize worlds, practices and conventions, and to function as a convincing argument for those who have not been in the laboratory. I have also argued that DNA fingerprint and matching likelihood number make an inseparable alliance in forensic cases, but that they transport different practices into court. The practice transported by the DNA fingerprint, familiarity with fingerprints as tools of identification, has proven to be questionable. With conventional fingerprints, all material can be used for identification (the whole print of all fingers, of one finger, half a finger, or even a vague print may do). For DNA evidence, one cannot "examine" all of the material.[55] Therefore, a selection is made based on variable regions on the DNA, the genetic markers. Categorizing DNA testing as a kind of fingerprinting suggests that one can read genetic information of each and every individual separately. This can be done in conventional fingerprinting. There, evidence can be identified (linked or unlinked to a suspect) if the fingerprint is included in "the population" of available fingerprints. The quest is then for the one and only match.

In DNA fingerprinting, establishing identification requires the exclusion of a match between the fingerprint of the evidence DNA and that of any other member in the control population and especially in the population at large.[56] Therefore, whereas conventional fingerprinting includes the biological trace in the population, DNA fingerprinting seeks to exclude it from the population. The difference between looking for other matches in the population and excluding or reducing the chance of a match in the population is paramount and may be difficult to overcome in DNA fingerprinting, as the example of the Turkish case has shown.[57] Whereas the conventional fingerprint gives a clear Yes or No – identity or no identity – the DNA fingerprint is based on frequencies and comes with a probability number.

Paul Rabinow also addressed this analogy.[58] He shows the irony in the promises made by both conventional and DNA fingerprinting. The British eugenicist and founding father of the fingerprint, Francis Galton, studied the fingerprint in the hope of developing a tool of classification between populations. In this he did not succeed.[59] The fingerprint showed no population structure. Instead, it established its prestige in forensics as a tool of individual identification. The DNA fingerprint has been developed and introduced into forensics as an ultimate tool of individual identification; however, as has become clear, individuality cannot be determined without situating the individual in a population. "Galton's regret" is indeed the weak spot in DNA evidence.

In this case, the fingerprint analogy seems to have lost ground and the question is whether DNA fingerprinting has ceased to be an immutable mobile. The matching likelihood number was instructive of the differences between both sides of the analogy. Matching likelihood computations blurred the fingerprint analogy and put the burden of proof on the markers typed, the reference population used and the allele frequencies presupposed. Yet the matching likelihood number determined the fate of the DNA fingerprint. As we have seen in the Turkish case, DNA fingerprints and especially matching likelihood numbers were travelling back and forth between laboratory and court room. Both DNA fingerprint and matching likelihood number proved to be *mutable*.[60] Their inscriptions and the significance of the practices and information carried through them changed several times. In addition, neither the courtroom nor the laboratory remained unchanged. Laboratory practice had been transported into court, and courtroom practices into the laboratory. Among other things, laboratory reports, control populations, different DNA profiles and methods of computation entered the door of the courtroom. A Turkish suspect, a Dutch control population and various Turkish populations found their way to the laboratory. Latour argued that the power of immutable mobiles is dependent on their ability to "recombine" different practices.[61] Can the matching likelihood number, the number-one immutable mobile, combine all these different worlds/practices and present them at once? In other words, is it capable of bringing the second immutable mobile, the DNA fingerprint, back to the courtroom?

In the following section we will view how Lab F enabled the DNA fingerprint to be brought back to the courtroom. We will see that in order to achieve this, the Turkish suspect had to be made into a T-suspect once again.

Back to the laboratory: making similarities

Given the objections and questions of the defence in the Turkish case, the DNA evidence seemed to be at risk. In the laboratory, different matching likelihood numbers were produced based on a set of markers of Lab F,[62] on the set of markers of the Laboratory for Criminal Justice and on the data of the German study, all of which produced figures around 10^{-6}. This figure did not seem to convince Lab F. Not only was the matching likelihood number larger than 10^{-7}, but also the set of markers in each comparison declined as a result of trying to use comparable markers.

There seemed to be no way round the problem of a suitable control population until another scientific paper appeared to show a way out of this stalemate situation and to help to take the DNA evidence back to court. The paper suggested a method for blurring the specificity of population.[63] Its authors had compared individual profiles with different reference populations, leading to the conclusion that allele frequencies may vary between populations depending on which marker is used. It was argued that the errors that occur when determining the DNA profile of an individual from a population other than the reference population could be reduced by using a particular statistical model. In addition, the paper suggested, the ties between an individual and a population are loosened if more genetic markers are typed.[64]

In the beginning of the Turkish case, it was stated that the DNA fingerprint produced in Lab F was based on seven genetic markers. With these markers, it was possible to produce a DNA profile and to compare it with that of another individual (as in the case of evidence and suspect DNA). This information did not lead to identification; it could not tell who these individuals were. The DNA evidence was inhibited and the profile did not become a fingerprint. For that a population was needed. Excluding a match with the rest of the population was not possible without having access to the right control population. Consequently, the individuality of the profile remained obscure. One could say that the problem of Lab F seemed to be with the population: the absence of an appropriate control population. As we will see below, the solution was, however, sought on the side of individuality. Since the problem raised by the defence was deemed plausible, and since the laboratory itself did not have access to Turkish samples, the way out of the impasse headed in another direction and the solution was laid in the hands of technology. When I asked the head of the laboratory to explain the meaning of the paper concerned, he said:

> If one compares two brothers on the basis of a single marker, the chance of a match is 50 percent. But when using 25 markers, the chance is 3×10^{-8}. An arbitrary comparison between any two individuals based on 25 markers gives a matching probability equal to zero. So, generally, the more markers one uses the smaller the chance that two individuals will look alike.

In terms of our case, the solution was to make the profile more individual by using more genetic markers. Instead of seven markers, 10 markers were used. In a way, this is a matter of statistics: the more variables one introduces, the more specific the units become. This refinement, using more genetic markers, made the profile of the Turkish suspect less population specific. In a sense, this made the Turkish suspect into a T-suspect, who had thus become a member of a much larger population; for the very reason that all members of that population look less alike and had become more individualized. Also the Dutch control population had become representative of a much larger population than the Dutch. The problem of "representativity" was resolved, because the control population of Lab F had become more sensitive, since all profiles had become more individual. The laboratory was now in a position to calculate the matching probability of "the Turkish suspect" by comparing his profile with those of the "Dutch control population." Based on this comparison, Lab F found a matching likelihood number of 10^{-10}. The DNA fingerprint produced was no longer the fingerprint of the *Turkish* suspect but that of a *T-suspect*. Because of the number of markers, the DNA profile of this individual could become evidence, since it could be expressed in a population. The suspect had thus become similar enough to be identified as different from the rest of the population.

Similarities established

The seventh concept of population is now introduced into the case. Earlier it was argued that forensics works under the presupposition that the suspect is innocent and that the perpetrator is out there in the population. The task is to determine the individuality of the suspect's profile by simulating a comparison between the individual and all other members of the population. Therefore, the suspect should be set apart in order to be sure that the specific combination of alleles (which make up the DNA profile) is unique and does not occur in the population. However, to achieve this, the suspect should also be sufficiently similar to the control population to help to estimate this probability. Without the presupposition of "similarity" (i.e. that the genetic profile of the suspect is represented in the population), identifying him proved to be impossible. The very presupposition of similarities and the objections to it have already revealed six different concepts of population. As we saw above, population might be a product of family names, of laboratory practice and routines, or of genetic proximity and distance. It could also be a product of race, national boundaries or of genetic markers and their specific clustering in different population. This makes clear the "problems," or rather the variety, in the practice of *population*. Lab F, however, sought a solution on the side of individuality. By introducing more markers into the analyses, both the profiles of the laboratory's control population as well as those of evidence and suspect became more individualized. This suggests a new feature of genetic markers, namely their number. First of all, more markers can distinguish better between individuals. The discriminating power of a marker is dependent on how many more markers can be used in a specific job at the same time. Second, the number of markers is crucial in producing differences or similarities between populations. When discussing the *sixth concept of population*, it was shown that the set of markers used did produce differences between what should be viewed as Turkish and what as Dutch. That distinction was based on fewer than five markers and, as we saw above, it was not countered with a set of seven markers. Using 10 markers, however, blurred that distinction and incorporated a new concept of population assisted by technology. Ten markers were capable of producing a population in which both Turkish and Dutch could fit.

The concept of population we have here is based on genetic markers and, more specifically, on their number. The more markers, the larger the population becomes and from the interview quoted it became clear that all the individuals in the world become part of one population when 25 markers are used.

Reporting on immutable mobiles

The case discussed in this chapter has produced many confusions. It started out with the issue of making individuality and became an issue of making population. Identification started as a matter of DNA and became that of technology. Meanwhile a T-suspect became Turkish and a T-suspect once again, while a control population became Dutch and then a control population once more.

There was a laboratory practice and a law practice and two mobiles inbetween: the matching likelihood number and the DNA fingerprint. These were the very devices that could make laboratory facts into court evidence and could make a link between these practices. Yet when mobilizing one practice to another, their immutability was at stake.

Whereas the focus of the court is primarily on individuals and on their identification, that of the laboratory is on variability in the DNA. However, in the interacting practice of law and science, the laboratory's concern is to produce individualized DNA. To do so, DNA is not treated in isolation but as both part and product of the socio-material network of DNA evidence. The very existence of this evidence could be seen as the work of laboratories and the handling of DNA. However, without the juristic regulation of laboratory practice, or without taking into account the specificity of the case at issue, a DNA profile may fail in a law practice. In the traffic between laboratory and courtroom, the DNA fingerprint and the matching likelihood number are the centerpieces of DNA evidence. They embody the interacting practices of law and science and usually express a "smooth" translation of scientific facts into court evidence. This was the matter of concern in the Turkish case discussed here. It began as a routine case, the suspect's profile being compared with the population as usual. Since the victim and the initial two suspects were all of Turkish descent, the defence started to question the laboratory's routine. From this, the practice of producing DNA evidence, and the practice of producing individuals and populations, was opened up for investigation. Whereas the DNA fingerprint had to testify to the individuality of the suspect, the matching likelihood number had to do so to the exclusion of other members of the population. Various versions of *population* have been produced to establish the individuality of the suspect and the evidence on the basis of DNA. Population may be a matter of family names, a matter of laboratory routine or one of genetic proximity and distance. Then again, it may be a product of racial differences and similarities, of national boundaries, of genetic markers or of the number of markers applied. These different versions do not add up to produce an integral picture of population. They are not pieces that can complete a puzzle. One reason for this is that any version of population embodies a specific set of techniques and materials and the treatment of these. Population embodies specific kinds of practice. The transportability of such practices (for example between laboratories) may be feasible, but it may also be inhibited. For example, a version of population as national boundaries, such as was used in the German study, required specific markers that were not part of Lab F's practice. A population based on family names may not be feasible in other contexts, because the social order implied in names may vary in different parts of the world, or because the routine of sampling may be a product of divergent practices, for example, one in which names are not recorded. Also in a context of genetic markers, population could no longer be practiced as race since genetic markers structure groups of individuals in ways that conflict with racial categorizations. The last hints at the second reason why the different versions of population identified here do not add up to one another. These versions may conflict, for they require different technologies for producing them and they simply produce different objects. For example, population based on national boundaries may conflict with one based on language, a source of political discomfort in various parts of the world.

Should this lead to the conclusion that, since population can be made into different things, this variety is a random collection consisting of autonomous elements? The examination of the Turkish case points in a different direction. It shows that differences between versions of population are neither natural nor essential but are matters of practice. It may be possible to make two conflicting versions fit. This, however, would require a change in laboratory practice: it would require additional technical interventions. Identifying the Turkish suspect also involved changes in practice; some of which were feasible, others not. For example, Lab F neither had access to the samples of the German study, which embodied a practice of national boundaries, nor could it link the data of that study to its own practice of seven genetic markers. However, a link between a population based on family names (the laboratory's control population) and an individual of Turkish descent was enabled by using more than seven genetic markers. This intervention accommodated a population in which Dutch and Turks could fit. Hence, population does not exist by itself but is enabled in specific practices and inhibited in others.

This treatment of population and of the practices in which it emerges suggests that (im)mutability and mobility are not issues restricted to the traffic of "things" between different domains, such as the domain of science and that of the law. The (im)mutability and mobility of scientific facts may also be at stake in the exchange between laboratories. Thus their fate also expresses the tension between similarities and differences between scientific practices.

Conclusions

In genetics, it is often claimed that individuals can be known by their DNA. In this chapter, I have shown that more is involved to achieve that. In genetic practice, neither the individual nor the population is a *natural* category. Both are technology assisted and established in the diverse practices of laboratory routines. I have examined how that is done.

In Chapter 1, we encountered the debate on defining population in the context of the Diversity Project. For the purpose of sampling, studying and analyzing differences between populations, various possibilities of defining population were at issue. As a result of these debates between geneticists, the consensus is to *define* population according to linguistic criteria. By way of contrast, my main concern here was to examine how populations are *enacted* in laboratory routines. Based on one particular case, I have shown that at least as many as seven different versions of population may be found in forensics and in population genetics at large. In a laboratory setting, neither the individual nor the population is treated as a matter of definition, but rather as a matter of technologies, established practices and routines. The conduct of scientific practice is heterogeneous, and the handling of scientific objects requires various different technologies, contributing to a diversity in what these objects "are." The question prompted by this concerning population, is of course, how do we want to be made into population?

Notes

1. The revision of 1994 is actually an addition to the Dutch Criminal Code (articles 151a, 195a, 195b, and 195d; see Kosto, 1994). A short description of forensic science in the Netherlands can be found in Jacobs (1995: 107–111). On expert witnessing in American lawsuits, see Jasanoff (1995). For a comparative study on expert witnessing in the Netherlands and the USA, see van Kampen (1998). Several other discussions, prompted by the well-known O. J. Simpson trial, occur in a special issue of *Social Studies of Science*, edited by Sheila Jasanoff and Michael Lynch (1998).
2. According to the Dutch Criminal Code (Wetboek van Strafvordering, articles 151aj, 195a), only laboratories assigned via "de algemene maatregel van bestuur," the political and judicial arrangement for forensic science, may produce DNA evidence. In the Netherlands, there are two of these: the Laboratory for Criminal Justice (Rijswijk), and the Forensic Laboratory for DNA Research (Leiden).
3. On rituals of scientific practice, see Jordan and Lynch (1992).
4. For the notion of science as a cultural practice, see, for example, Pickering (1992).
5. I learned that laboratories do not necessarily have to be the immured sanctuaries, the domains of no entrance for outsiders, as described by Bruno Latour, for example. When I started my studies, I had several talks with Professor Gert-Jan van Ommen, head of the Department of Human Genetics, to which the forensic laboratory is linked, and Vice-President of HUGO. He advised me to visit one of the Diversity Project conferences organized by HUGO (in Barcelona, November 1995) to get to know that particular scientific community. Since I was particularly interested in daily laboratory routine, he suggested that I talk to Dr Peter de Knijff, head of the forensic laboratory, who was also working on human genetic diversity. Dr de Knijff was enthusiastic about an outsider interested in genetic practices and

was willing to arrange training for me in his laboratory. Having attended the conference in Barcelona and other international meetings, I learned, to my surprise, that many laboratories participating in the Diversity Project were actually very open to outsiders. In the two where I conducted my research, I combined a "hands-on" project (training or a small genetic research project) with ethnographic work. Consequently, I kept two journals: one laboratory journal on which I worked in the laboratory and an ethnographic journal, which I often had to write at home. At the end of my laboratory research, I conducted interviews with laboratory members.

6. Polymerase chain reaction (PCR) was the revolutionary cloning technology of the late 1980s. It is a DNA fragment copying technology using a thermostable enzyme. This procedure not only produces more DNA, making it easier to study, but it also allows a chemical group (such as a fluorescent group) to be attached to the copies so that they can be detected using ultraviolet or laser beams. PCR mimics a "natural process" that takes place in the cell, namely DNA replication, allowing geneticists to copy a target fragment of DNA in a short time. The enormous number of copies produced by PCR enables further studies on the DNA fragment. For an exciting account of the different actors involved in the making of PCR, see Rabinow (1996a). On the adaptation of PCR to fit the various needs of scientific worlds, see Jordan and Lynch (1998).

7. For laboratory-specific typology and the ethnographic practice it may embody, see Mol (2000).

8. NRC (1996). This committee was installed by NRC in 1989 and had issued a previous report (NRC, 1992). The 1992 report regulated the use of DNA evidence in forensic cases but was also seen as highly critical of this method, contributing to controversies around evidence DNA. William Sessions, a former director of the US Federal Bureau of Investigation, asked the NRC to initiate a follow-up study in order to clarify the controversial character of DNA evidence.

9. NRC (1996: 82). Since in the Netherlands only laboratories accredited by the Board of Accreditation are allowed to produce DNA evidence, elaborate protocols describing each step of the experiments and taking into account the risk of contamination were already operative in 1994. For almost all other laboratories in Europe (including Germany and the UK) as well as in the USA, such protocols did not exist at the time, and the NRC report became a guideline for forensic practices. Nevertheless, in the Netherlands it also functioned as a point of reference, and it became especially directive for the application of statistical analyses and for calculating matching likelihood estimates in cases of population admixture. For the relevance of the NRC report in the domain of statistics, see the article by the statistician and the head of the Laboratory of Criminal Justice (Department of Forensic DNA) (Sjerps and Kloosterman, 1999).

10. The Board of Accreditation is the highest institution that monitors and assesses the technological preconditions for forensic work. There are several of these boards in the Netherlands, but both the laboratories involved in evidence DNA are accredited by the strictest board, the so-called SterLab. Every year an audit takes place and all aspects – laboratory space, technologies, paperwork concerning how cases are reported and stored, and protocols of the laboratories – are inspected. Whereas this board is an evaluative institution, both laboratories also have a quality-control manager who supervises the daily work – that is, the laboratory space and protocols, and the conduct of laboratory members within this space (which tests should be done in flow cabinets, which under the "hut"

(fume-hood)) – and who makes sure that everyone obeys the clothing regulations (when to wear gloves and masks, laboratory coats, etc.).

11. Information about the DNA of the laboratory members is source of many practical jokes. Specific behavior of individuals is then jokingly linked to their genetic outlook: a very strong Y-chromosome and yet not so good a football player; or a strange peak that could indicate an extra X-chromosome in the case of a male member.

12. This differs from some other countries, such as the USA, where reports have to include an error rate for the results.

13. This is one of the possible understandings of genetic markers. Markers can best be seen as hybrids, being simultaneously objects of study, the technology to do that and the signs that confirm their presence; see Chapter 3 for an analysis of markers.

14. A *base-pair* consists of two nucleotides bonded together, one located on each of the two strands of the DNA molecule. Since the nucleotides are distinct chemical groups, the ways they can bond with each other are limited; the commonest possibilities are A–T (adenine–thymine) and C–G (cytosine–guanine).

15. The poly-markers are: LDR, GYPA, HBGG D7S8 and GC. Some of these are in coding DNA; others are in introns (the flanking regions of a gene) and so at present are not known to be involved in vital functions of the cell. Note that these markers are not based on variation in fragment length but on "molecular weight," a substitution of a base-pair by another. Since the different nucleotides consist of different chemical groups, they differ in weight.

16. NRC (1996: 26). Although, in the case as a whole, other evidence may reduce the chance of this probability, the laboratory is tied to this working hypothesis by the 1994 law in the Netherlands. One could say that the conduct of law and science leads to another type of evidence, sought in laboratory practice rather than courtroom practice. See also Sheila Jasanoff (1995: 49–68), who addressed this difference and analyzed what counts as fact in court and in science.

17. This figure may differ depending on the judge conducting the trial: some accept a higher matching likelihood number while others demand a much lower matching figure.

18. For a clear elaboration of the difference between *inclusion* and *identification*, see Rabinow (1993: 60) (also published in Rabinow (1996b)); Lander (1992: 191–193). Lander's contribution gives a comprehensible overview of how DNA typing is done in forensics and addresses some controversial cases in the USA and their implications for future forensic work.

19. Since I did not talk to the defence myself, I introduce this example to show that the case did not stand alone but linked up with many debates and controversies surrounding evidence DNA outside the Netherlands.

20. Lewontin (1993). My reference is to the Dutch edition (Lewontin, 1995b: 99–100). Another well-known forensic case, the "Castro Case," became a controversy in the USA because of sloppy conduct in the laboratory; see Lander (1992: 196–201) and Jasanoff (1995: 55–57).

21. Latour (1990: 56), second emphasis added. See also Latour (1987), especially Chapter 6.

22. Latour (1990: 26 and 44–47).

23. An analogy can, therefore, be seen as an apparatus of signification (Strathern, 1995a: 13). For a straightforward and motivated use of this analogy, see NRC (1996: 14). For an example of this analogy in practice, see Jeffreys *et al.* (1991).

For a study of the use of (conventional) fingerprinting at the beginning of the twentieth century, see Cole (1998).

24. Rabinow (1993: 60).

25. Rabinow (1993: 63).

26. For the convenience of scientific prestige and its importance for the credibility of expert witnesses in court cases, see Cole (1998). In his elegant paper, Cole argues that the "ordinariness" or the familiarity of the fingerprint, in fact, almost jeopardized its testimonial power in lawsuits. Its role in court was based on a delicate balance between strangeness, attending to a scientific practice, and familiarity, of which the validity could be observed in the courtroom. One could say that this double role in the case of DNA evidence is nicely delegated to the two components: the DNA profile and the matching likelihood number.

27. Note that the Department of Human Genetics is part of the laboratory network. Moreover, it is the so-called pen-wielder of the Forensic Laboratory, through which it can be treated as part of Leiden University.

28. TNO (Nederlandse Organizatie voor Toegepast-Natuurwetenschappelijk Onderzoek) has a close tie to the TNO-Leiden, where the head of the laboratory used to hold a position conducting and guiding research on heart disease.

29. This type of sampling is called convenience sampling, as opposed to simple sampling. The latter is based on a random sampling procedure, whereas the former is based on a random selection of already existing samples (from blood banks, paternity testing or laboratory personnel) (NRC, 1996: 30 and 186). For further information about the male samples, see de Knijff (1992: 57); Roewer *et al.* (1996: 1032).

30. In an article written in collaboration with the Forensic Laboratory for DNA Research, the following is stated: "The two groups of unrelated males analysed in this study (70 Germans and 89 Dutch) comprised controls routinely used for the validation of forensic genetic markers. Care was taken that none of the males share last names, and that all were white Caucasians" (Roewer *et al.*, 1996: 1032).

31. Whereas before 1960 most immigrants in the Netherlands came from former Dutch colonies (such as Indonesia and Surinam) or from southern European countries, since the 1960s a large group of (male) immigrants have been recruited in Turkey and Morocco by government officials to counter the pressure on the Dutch labor market. The recruitment initiated a much larger migration wave from these countries, especially through family reunion. Of the two million immigrants living in the Netherlands nowadays, around 240 000 are of Turkish descent and live in high concentrations in large cities (Metze, 1996). Despite the fact that this group of immigrants has been part of Dutch society for almost four decades, they often appear to be "a special group" or "an outgroup" (Metze, 1996: 32).

32. Compare Harold Garfinkel (1996b: 191), where he addresses routine practices embodied in the records of an outpatient psychiatric clinic. He argues that these records may cause "normal, natural troubles" for a sociologist who uses them for research, because the practices they embody are not made explicit and tend to escape the eye: "'Normal, natural' troubles are troubles that occur because clinic persons have established ways of reporting their activities; because clinic persons as self-reporters comply with these established ways; and because the reporting system and reporter's self-reporting activities are integral features of the clinic's *usual* [emphasis added] ways of getting each day's work done – ways that for clinic persons are right ways."

33. For this notion of reflexivity, see Garfinkel (1996); Lynch (1997). For a comprehensive elaboration of the difference between *reflexivity* in the approach of ethnomethodology and *self-reflexivity* in other sociological approaches, see Lynch (1997: 34–39).
34. On the clustering of alleles according to population, see HUGO (1993: 24).
35. These two publications were based on collaborations between scientists in Turkey and Germany (Alper *et al.*, 1995a,b). The third study is Meneva and Ülküer (1995).
36. Dr Peter de Knijff, interview with author at the Forensic Laboratory, Leiden, 17 January 1997.
37. Dr Peter de Knijff, personal communication.
38. For example, in November 1995 in Barcelona at the international conference *Human Genome Variation in Europe: DNA Markers*, a special plenary session was held on genetic markers for typing genetic variation (Bertranpetit, 1995). The main goal of the discussion was to develop a set of "priority markers" that could be used within the realm of the Diversity Project. As a result, a preliminary document was produced in February 1996 by Jaume Bertranpetit, in which these markers were described. The document was circulated as a personal communication.
39. This discord is best seen in the publications of Richard Lewontin and Daniel Hartl (1991) on the one hand, and Rannajit Chakraborty and Kenneth Kidd on the other (1991). For an account of the controversy around these papers in *Science*, see Lewontin (1993: Chapter 4). Furthermore, the US committee on DNA Forensic Science (see note 8), a co-production of the NRC and the Commission on DNA Forensic Science, can be seen as a capstone in this debate. For paternity DNA studies and DNA fingerprinting on the basis of minisatellites, see Jeffreys *et al.* (1991).
40. My focus in this analysis is on criteria and the more "technical" features of genetic markers as applied tools in forensics; see Chapter 3 for an elaboration.
41. NRC (1996: 14).
42. NRC (1996: 26).
43. Personal talks with members of the forensic laboratory. In cases where markers on the same chromosome are chosen, the loci should be far apart to allow for independent inheritance; these loci are said to be non-homologous. For an analysis of the concept of homology in the realm of the HGP, see Fujimura and Fortun (1996).
44. See NRC (1996: 34); furthermore it is suggested here that markers with a number of alleles lower than five should be rebinned (grouped in bins containing at least five alleles). Note that for paternity testing the number of alleles should preferably be lower than for evidence DNA profile typing – since higher variability indicates a higher mutation rate (in a locus); in these cases, even mutations from one generation to the other may occur. See, for example, a review article by Jobling and Tyler-Smith (1995).
45. Also in the German study (Alper *et al.*, 1995a: 93), the Turkish population in Turkey is referred to as Caucasian.
46. On Mendelian population see, for example, Macbeth (1993: 51 and 54). One striking example of how race "is done" in genetics can be found in Jeffreys *et al.* (1991: 825): "The only preselection of data for this study was that of ethnicity [Caucasian], which was determined on the basis of photographic evidence."
47. See also Duster (1992, 1996); Rabinow (1996b); Lock (2001).
48. An interesting feature of the use of racial taxonomies in population genetics can be found when comparing genetic discourse in the USA with that in Europe: in

Europe the main races are Caucasian, Negroid, and Mongoloid, whereas
taxonomies in the USA produce more races, such as Caucasian, Blacks, Hispanic,
East Asian, and American Indian. For an example, see NRC (1996: 35). On the
different taxonomies of race in genetics, see Duster (1996).

49. See NRC (1996: 34ff.).
50. For a history of race, crime and the law, which takes the social construction of both
    crime and race into account, see Duster (1992). On the intertwined history of race
    and genetics, see Kevles (1985, 1992); Chapman (1993). For a historical debate on
    biology and the human races after the Second World War, see UNESCO (1952).
    For an analysis of this document, see Haraway (1992: 197–203). A
    broad collection of papers on race and the sciences is edited by Harding
    (1993).
51. The collision of population and race becomes clear, for example, in the following
    quotation (Dunn, 1951: 24): "Since biologically *races* are *populations* differing in
    the relative frequencies of some of their genes, the four factors noted above
    [mutation, selection, genetic drift, and migration/mixing] as those which upset the
    equilibrium and change the frequencies of genes are the chief biological process
    responsible for race formation" (emphasis added).
52. Alper *et al.* (1995b: 112), emphasis added.
53. See Macbeth (1993: 49, 78ff.) about national and population boundaries (the
    former being referred to as a "conceptual boundary"). She suggested national
    boundaries as one possible approach to compare populations, since these
    boundaries often coincide with other "natural" boundaries. The problems of this
    perspective are of course clear if one looks at, for example, the map of Africa. For
    a conception of differences within and between populations, see, for example,
    Chakraborty and Kidd (1991: 1737). Moreover, the statement about differences
    within and between population is also used when other than national boundaries
    are seen as criteria of difference between populations. For a critique of and an
    elaboration on this argument see, for example, Lewontin and Hartle
    (1991).
54. On population admixture, the clustering of markers within populations and
    subpopulations, and the calculation of matching likelihood probabilities, see
    Lander (1992: 205); Lewontin and Hartl (1991: 1746).
55. Lewontin and Hartl (1991: 1746) argued that the analogy does not hold water if
    one takes into account the material that can be studied in evidence DNA (only a
    fraction of the retrieved DNA is used), and the technology (the small number of
    markers that were available in the early 1990s). Their argument is that DNA
    fingerprints do not contain as much information as conventional fingerprints. In a
    personal communication, Richard Lewontin made it clear that DNA profile typing
    has become more powerful thanks to more and convincing genetic markers
    (discussed at the *fifth Annual Meeting of the Society for Molecular Biology and
    Evolution*, Garmisch-Partenkirchen, Germany, June 1997).
56. By the late 1990s, the amount of information stored in large databanks had grown
    dramatically, making it possible to look for matches between DNA profiles,
    especially if suspects have a criminal record. On the history of race and the
    incrimination of (groups of) individuals, see Duster (1992, 1996).
57. I thank Dr Hans Zichler of the Laboratory for Evolution and Human Genetics,
    Munich, for having brought this point to my attention and clarified my thoughts
    about the analogy.
58. Rabinow (1993).
59. Galton (1892).

60. See also Latour (1987), who has discussed that mobiles cannot be entirely immutable.
61. Latour (1990: 45).
62. The set of markers used at this stage is smaller and consists of STRs and HLA markers only.
63. Chakraborty *et al.* (1993).
64. This argument is put forward in the following words: ". . . in general, the profile frequency is a decreasing function of the number of loci scored" (Chakraborty *et al.*, 1993: 68).

3

Ten chimpanzees in a laboratory: how a human genetic marker may become a good genetic marker for typing chimpanzees

Introducing the argument

This chapter deals with genetic markers. In the field of population genetics, markers are crucial categories. They are the very objects of comparison between individuals or between populations. This has become clear from Chapter 2. The question in this chapter is, therefore, what *is* a genetic marker? To answer this, I will treat a marker neither as a quality embodied in the DNA nor as an autonomous category that can be investigated everywhere. Rather, I will "study around it" and examine the socio-technical network of laboratory routines in which it is enacted. Markers, as will become clear, are hybrids, which involve DNA, technologies and ways of aligning these. The argument put forward is that genetic markers are technically and locally invested, and that this quality co-determines their ability to move from one locale to another.

Genetic markers are often presented as innocent tools, as *loci* present on the DNA that need only the keen eye of technology to make them emerge. Population geneticists have become increasingly aware of the lack of universality of these tools and of their embeddedness in different populations as well as in different laboratory practices. Nevertheless, the dream of genetics is to find universal markers, through trial and error or through large-scale studies. This is, in a way, a quest for the unproblematic tool that will make it possible to focus more on populations and less on the technology at hand. One could call this dream the quest for an "unbiased eye" that can see without regard to populations or individuals. The dream of *good* genetic markers, so it seems, is nowhere and everywhere at the same time.

This chapter will address the tension between the local and the global qualities of genetic markers by taking seriously both the practicalities of markers – as tools of everyday laboratory practice and the need of scientists to work together – developing standardized or *good* genetic markers. The typing of chimpanzee DNA is studied in order to show that the DNA fragment, the technical means to visualize that fragment, and the goals for studying DNA all become constituent parts of a genetic marker. These constituent parts will together determine what a marker is and what may count as a *good* genetic marker.

Markers: a round-table discussion

Let me take you to a round-table discussion that took place after three days of conferring on human genome diversity.[1] The theme was genetic markers. To quote some remarks made by a number of participants.[2]

"A list of markers should be made, as an indication to newcomers in the field and in order to compare the different data." (Jaume Bertranpetit).

"We need to have some consent about the markers so as to compare the results." (Luca Cavalli-Sforza).

"Criteria for markers should be that they show variation between populations." (Luca Cavalli-Sforza).

"We need markers that are selectively neutral to different population structures." (Sir Walter Bodmer).

"Preferably markers that do not require use of radioactivity." (Svante Pääbo).

"What we need right now is a list of priority markers." (Jaume Bertranpetit).

"What is the use of such a list if people aren't working with it anyway?" (Brian Sykes).

"One could recommend things now, but it would be preferable to choose a democratic procedure, such as people reacting from their own experience of research." (Svante Pääbo).

"What are good genetic markers?" (Svante Pääbo).

We will return to these points at the end of the chapter. The discussion was part of the conference *Human Genome Variation in Europe: DNA Markers* held in Barcelona in 1995. The conference was aimed at fine-tuning a variety of laboratory practices, scientific goals and criteria for population genetic research, by paying special attention to an important issue in this field, namely genetic markers.[3] A "list of priority markers" should do the job of fine-tuning.[4] Particularly the last three remarks quoted indicate that this is not an easy job. The round-table discussion thus revealed a tension between various local needs, interests, goals and practices. The question raised by one of the participants, namely "What are *good* genetic markers?" hints at a variety of practices and suggests that markers bear this tension as well. This question refers us to the heterogeneity of scientific work where nature and technology are aligned.

In line with various ethnographies of laboratory science, I will examine how this alignment is established, and what kind of components it contains.[5] I will focus on how markers are 'enacted' in laboratory routines.[6] So the main aim of this chapter is what is it made to be in such a setting. Moreover, taking up on issues raised in the round-table discussion, I will also view how markers may become *good* genetic markers, and how the tension between a "standardized" global approach to genetic diversity and the heterogeneity of laboratory work is handled in such a setting. In other words, we will examine the normative content of genetic markers as "standards": when they become part of laboratory routines and start to function as *good* genetic markers. As the round-table discussion indicated, scientists do not work in isolation. A major part of their work consists of traffic in objects, technologies and knowledge, and of the facilitation of such traffic.[7] In order to establish collaborations and comparisons between work conducted in different time and places, standards are crucial.[8] Standards, however, tend to obscure their normative content. Yet, the practices involved and the content embodied open up for investigation when routines are disturbed, for example when the DNA of chimpanzees became the object of research in a laboratory for human genetics, as discussed below.[9] Before we enter the laboratory, let us first take a look at a definition of a marker.

Markers: a definition

Marker: an identifiable physical location on a chromosome whose inheritance can be monitored. Markers can be expressed regions of DNA (genes), a sequence of bases that can be identified by restriction enzymes, or a segment of DNA with no known coding function but whose pattern of inheritance can be determined . . .[10]

So a marker is a specified fragment of the DNA inherited *unchanged* from one individual by another. Such fragments can be identified by their "physical location" on the DNA and can be monitored by geneticists.[11] This definition emphasizes the relevance of markers in terms of inheritance and indicates that markers are objects of comparison. They cannot be studied in one individual but only in more than one by comparing them. Conversely, one could say that in genetics individuals are not so much related by blood or by DNA for that matter. Rather they are related by genetic markers. Therefore, markers are the stuff that genetic lineage is made of.

Not the DNA but a marker

Geneticists do not study the whole DNA (the so-called genome) of an individual or a population, but only small fragments. The DNA molecule of a human contains too much information for that. Consider the fact that geneticists from different parts of the world have been constructing and mapping *one* human genome ever since 1989 and had "finished" doing so in February 2001 through the publication of a "draft sequence" in the journals *Nature* and *Science*.[12] To make the picture complete, consider also the fact that what is often referred to as the *genome* is the so-called coding region and makes up only 5% or less of the whole DNA molecule.[13] The non-coding region is often referred to as "junk DNA" and is hardly ever considered as part of the genome. Therefore, scientists can only study small fragments of the molecule, either coding or non-coding DNA, defined by their interests and research questions. These fragments may be referred to as markers. Hence not the DNA, but genetic markers are objects of study.

The first reading of the definition suggests that in order to know what a marker is we should look at nature. A marker is a DNA fragment with specific qualities. However, a closer look at it already suggests an intertwinement between nature and scientific practice. In the definition, it is stated that a genetic marker is both the *information* encoded in the DNA, a "gene" or "some segment of DNA," and also its "physical location," namely a specified sequence and its location in the DNA molecule. Whereas the first is concerned with the functioning of DNA, namely *how* genes play a role or *how* a DNA fragment is inherited, the second is concerned with *where* the sequence is and *what* it looks like in terms of nucleotide order.[14] The definition embodies both accounts, but there seems to be a primacy of the first over the second when the importance of monitoring is being considered. A genetic marker is defined as "a physical location . . . whose inheritance *can* be monitored." Therefore, the criterion of monitoring, or the feasibility of studying its patterns of inheritance, seems to be a precondition for a DNA fragment to become a marker. On this basis, a marker is not only about nature out there. Rather, it implies a fit between a natural process and a scientific practice in which it can be studied and in which it can be made into an object of research.

How this is done is at the heart of the case studied in this chapter. In the following, we will leave the realm of definitions and turn our attention to practices. There we will meet geneticists who strive to answer the question whether *good* human genetic markers (i.e. markers that have become standards in studies of genetic variations in human populations) could be applied in the diversity studies of chimpanzees. We will follow the geneticists' investigations, take into account the technicalities that never make it into the handbooks and focus on the intertwinement of DNA and technologies in such practices. Let us for this purpose again enter Lab F.

Laboratory practice

The first day in the laboratory

On Monday March 18 1996, the rail connection between Amsterdam Central Station and Leiden Station was bad. That morning, the Amsterdam station was a scene of people running from one platform to the other while trying to listen to the information coming through the speakers about their next possible connection. As a result of this chaos, I arrived late on my first day in the laboratory. Having made my way through a labyrinth of corridors, I was surprised to be welcomed by a group that had learned about my delay by listening to the radio.

My trip to Leiden was not without preparation. I had asked Lab F for a short training course in some of the basic tasks of a technician. After introducing me to the laboratory members, the head appointed a daily supervisor for me, explained the project I was going to work on and promised that before the end of the day I would have done my first DNA extraction. Indeed, in the afternoon we were extracting DNA from blood. Before I knew it, bloodspots that belonged to 10 male chimpanzees known as Fauzi, Carl, Yoran, Zorro and their mates had been changed into DNA samples marked as TNO-CH1, TNO-CH2, TNO-CH3, TNO-CH4 and so forth.

But hold on! "Wasn't DNA supposed to look white?" I asked my supervisor. I explained that I had seen Kenneth Kidd (a population geneticist) on TV, demonstrating a white wool-like substance to the viewer. What we had was a clear solution instead. He told me that we were working with small amounts of blood and could not extract that much DNA from them. However the tiny bits of DNA would be sufficient because we would be able to copy them, using the PCR machines. "Ha," I said, "the polymerase chain reaction, the Nobel Prize-winning cloning technology." We both tried to recall the name of its inventor, Kary Mullis. We placed the rack with the labeled cups containing DNA in the refrigerator and left the so-called "pre-lab" (see Chapter 2).

Markers in laboratory practice

On my second day in the laboratory, I had my first encounter with a genetic marker. We ran a PCR to test one of the human genetic markers, *DYS 389 I–II*, on the chimpanzee DNA. Even with the help of a protocol and a supervisor, "setting up" a PCR for the first time proved to be a complex procedure. It required using and distinguishing between three different pipettes, which look the same but pipette different volumes; distinguishing between the different chemicals; pipetting and mixing in the right order and adding different solutions to a specified volume of DNA. Undivided attention was crucial here, because you had to do this for various individuals simultaneously. With such small volumes, it was easy to make mistakes, such as forgetting to add a chemical to a sample or adding a chemical twice. After having prepared the samples and being instructed about the storage of the different ingredients (DNA samples and various chemicals, also called reagents), we moved from the pre-lab to the post-lab to load the samples into the PCR machine.

While the PCR was running, my supervisor and I had a talk about what was happening to the DNA in there. He made some drawings to explain it. It became clear that the copying of DNA during PCR was a mimicking of nature. A prefabricated enzyme (a thermostable DNA polymerase), which was part of the solution, assisted this process. During a time-designated process of heating and cooling (the so-called cycles), the double-stranded DNA is first straightened and pulled apart (the so-called denaturing of DNA) and then copied before it clings back together. The copies complement the single strands and are produced by using DNA building blocks (nucleotides) and elongated by the polymerase, all added to the solution. Most important, not the whole DNA but only the marker fragment is copied. This is because *primers* are used: short synthesized sequences matching the beginning and the end of the marker fragment. By attaching themselves to the target sequence, they expedite the copying of that specific part of the sequence. In a way, the primers come between the two pre-existing DNA strands and prevent their clinging together. In the process of becoming a double strand, the primers force the single strand to use the DNA building blocks and to produce a copy. Thus the single strand is forced to cling back using a copy instead of an existing single strand of the template DNA. The primers not only match the beginning and the end of the target fragment, they also mark this fragment. The primers are labeled with chemical groups (radioactive, fluorescent or biotin groups) that assist the visualization of the marker fragment after PCR. Moreover, the process of copying is not linear but exponential: after one PCR cycle, the double strands of such a fragment would be copied into four, after the next cycle, four would have become 16; 16 turns into 256, etc. Within less than two hours, there might be a million copies. If the amplification works as expected, the marker fragment is available in prodigious amounts at the end of the PCR run.[15]

When the run was finished, I followed the instructions of my supervisor and loaded the PCR products (the multiplied DNA fragments) onto an agarose gel for electrophoresis.[16] The agarose gel was placed in a bath containing a buffer (to keep the pH steady ethylenediaminetetraacetic acid (EDTA)) and had a number of slots into which the PCR products had to be loaded. Before this is done, the PCR products are taken up in a blue-coloured solution (the loading mix), which allows the migration of the samples over the gel to be seen once an electric current is applied. The gel itself plays a key role in the visualization of the DNA fragments. A chemical group (ethidium bromide) added to the solution before it sets into the gel is crucial to this process. Ethidium bromide binds to another additive in the PCR product. This possibility was already provided during DNA extraction, a so-called "chelex extraction." When I asked my supervisor about chelex, he suggested that I pictured it like tiny pellets installing themselves between the helix-shaped DNA. The presence of these pellets in the PCR products allows for a chemical binding between chelex and the ethidium bromide in the gel substance; this binding allows the location of a DNA fragment after electrophoresis to be found. The gel thus becomes part of the visualizing technology.

When loading the gel, the first slot of the gel is usually reserved for the ladder. The ladder is a synthesized compound of DNA fragments whose sizes are known, such as 50, 100, 150 base-pairs and so on. Because the ladder starts to "migrate" together with the rest of the samples, it helps to determine the fragment length of those samples.[17]

Once the samples had been loaded into the gel, the bath is covered and the current set to 60 V, the samples indeed started to move, leaving a faint blue trace behind. I was instructed to set the timer for 20 minutes. Consequently, we had to wait that long to undertake the next step so I went out of the laboratory to have a short break. On my return, I found that my supervisor had interrupted the run. He looked surprised and a bit embarrassed when I entered. He had been unable to wait the 20 minutes wanted to have a quick look at the results by exposing the gel to ultraviolet rays. At the same time he looked very excited and cried out: "They did it, it worked!" Since it was just my second day in the laboratory, I did not altogether understand what he was trying to say. A second technician had joined us and he started to point at the orange-coloured bands on the agarose gel, which had lit up under the ultraviolet rays.[18] I slowly understood the nature of their excitement, and only a few days later I did really understand the relevance of what we were looking at.

Markers: intertwining nature and technology

Given this account, what then is a genetic marker? From the marker definition introduced above, we have learned that markers are DNA fragments inherited by one individual from another: it was defined as the *object* of study. The encounter with markers in Lab F illustrates the technical procedures involved in typing them. A marker is not by itself. In order to be a marker, a DNA fragment had to be aligned to various technical procedures and components.[19] The template DNA had to be integrated in a technological system in which chemicals and gels, marking and copying of DNA, protocols and precise work were aligned. One could say that in order to produce a "selective perception," namely the study of a particular fragment of DNA, the DNA had to be "upgraded" so as to make it studiable.[20]

Moreover, the account presented above revealed that next to being the *object* of study, genetic markers had a second quality: they had become a technology.[21] From the first step of the experiment, crucial components of the visualizing technology were added to the DNA. Anticipating its visualization, the copied fragment had become inseparable from chemical groups (chelex and fluorescent) and consequently from other components, such as the agarose gel or the ultraviolet beams. This entanglement suggests that the marker is not only the *object* of visualization but also the very *technology* to do that. The intertwinement of objects and technology that I am suggesting here is not so much about the fact that different technologies reveal different aspects of a specific object. Rather my argument is that the studying of objects in laboratories requires that the objects themselves are transformed into the means to do that, namely into a technology. A marker is, therefore, a hybrid. It is both an object and the technology to confirm its presence.[22]

Having addressed the routines of visualization, let us now focus on the routines of producing genetic similarities and differences, the main aim of visualization. Let me for this purpose introduce the chimpanzee case. A case, as indicated above, that disturbed the daily practice of Lab F and which, therefore, allows for a closer look at standards and routines.

Ten chimpanzees in the laboratory

Now why was Lab F developing an interest in chimpanzees, and why is it that one could find them in Lab F?

Lab F had received blood samples from five different primate (10 species male samples from each primate population) from the Biomedical Primate Research Centre: the Dutch Primate Centre.[23] They had asked Lab F to explore the possibility of developing *genetic passports* for primates. The passports would be supplied as chips and inserted into the primates' bodies. In 1995, the Dutch government had announced its intention to "monitor" the international trade in primates and to restrict their import for scientific research as well as for zoos. The idea was to prohibit the import of primates into the Netherlands and to breed them as much as possible in the Dutch Primate Centre, especially primates that were meant for scientific research. It appeared to be common practice to mix up the identities of individual primates for financial or research reasons. Laboratories would claim to be experimenting with the same primate, when actually they would be experimenting with a second primate because the first had died. Identifying primates by their genes seemed to offer a solution to these problems. A genetic passport would also be of interest to the Primate Centre as a means of assessing the loyalty of their clients: to determine whether they would predominantly buy from them or also from other primate suppliers in Europe.

Requesting Lab F in particular to conduct a pilot study is not that strange. As the name of the laboratory indicates, Lab F is a forensic laboratory and has, therefore, developed an expertise in identifying human individuals in forensic cases.[24] Lab F supplies DNA evidence for the court, based on DNA analysis, that confirms whether or not two genetic profiles of suspect and evidence material coincide. This individuality-producing practice suggests that Lab F might be very appropriate to answer the question of the Dutch Primate Centre. If Lab F can identify human individuals, might it not have the expertise to identify non-human individuals? This project was also of interest to Lab F for at least two reasons. If the answer to the question put by the Primate Centre proved to be positive, Lab F would have a fair chance of being asked to produce the genetic passports for all the primates at the Primate Centre. The second reason is connected to the laboratory's field of research. As noted above, the Dutch Primate Centre sent only *male* samples for this pilot study. This choice is not self-evident given the goal of the Primate Centre, namely to develop genetic passports for *all* their primates, both male and female. Having just introduced a set of Y-chromosomal markers, Lab F had developed an interest in testing these markers for non-human primates.[25] Since the markers are located on the Y-chromosome, a male-specific chromosome, the pilot did not aim at studying male and female primates equally, and it was provisionally reduced to a *male* primates project.

To answer the question of the Dutch Primate Centre concerning the genetic identification of primates, Lab F scheduled the typing of the chimpanzees first. It had practical reasons for doing this. It was not altogether clear whether these human genetic markers would work in primates under the established laboratory conditions. And since chimpanzees are considered human's next of kin, it made sense to start there, where the fewest genetic differences were to be expected.[26]

Y-chromosomal markers

In order to understand Lab F's interest in Y-chromosomal markers, let us take a closer look at the markers themselves. In addition to forensic work, Lab F had also developed research projects in the field of population genetics. After attending a forensic meeting in Berlin in 1995, the head of Lab F established a joint project with another forensic laboratory in the former East Berlin.[27] The laboratory in Berlin had a set of Y-chromosomal markers and was seeking joint projects with other (non-German) forensic laboratories to test these markers further and to evaluate their use for population studies and forensics. A set of seven markers had thus made the journey from Berlin to Leiden and was primarily being studied using the databases of these two laboratories.[28] One of the results of the collaboration between Leiden and Berlin was a paper published in April 1996. Under the heading *Discussion*, it stated:

> We have demonstrated, for the first time, how sensitive PCR-based methods can be used to characterize highly informative haplotypes of Y-chromosomal microsatellite loci. With four out of seven microsatellites presented, samples of Y-chromosomes could readily be differentiated with respect to their Dutch or German origin on the basis of allele frequency alone . . . , and as many as 77 haplotypes have been observed for these loci among the 159 males tested.[29]

By comparing males in Germany and the Netherlands, two neighboring populations, and detecting variation between them, these scientists found a strong argument in favour of Y-chromosomal markers, described as *microsatellites* above (small DNA polymorphisms). According to the authors, Y-chromosomal markers are *good* genetic markers because they show differences *between* populations.[30] Furthermore, the paper also indicated that "the large within-population diversities noted for haplotypes of Y-chromosomal microsatellites will render them useful markers for forensic purposes."[31] Therefore, this set of Y-chromosomal markers would allow studies to be made not only of the genetic differences *between* closely related populations but also of differences *within* these populations.

This paper was in preparation when the head of Lab F was invited to the conference *Human Genome Variation Europe: DNA Markers*. There he gave a talk about Y-chromosomal markers and made a similar argument presenting results based on a comparison between Dutch and Inuit (Eskimo) populations.[32] Many geneticists showed an interest in these markers and a number of joint projects with other laboratories were initiated. The head of Lab F was also invited to contribute to a report about genetic markers based on the round-table discussion described earlier in this chapter. This report was circulated among the participants and the head of Lab F contributed a section on Y-chromosomal markers.[33] In early 1997, two more papers were published, co-written by 27 geneticists, reporting a large-scale study conducted in various laboratories and comparing Y-chromosomal markers for a large number of populations. Their use in the field of population genetics was recommended and conditions for forensic application were indicated.[34]

One could say that these Y-chromosomal markers have become good genetic markers. They have become standards because the practices they involve were transformed in stable forms, such as protocols and scientific papers, allowing for their *mobility* between laboratories.[35] They have travelled from Berlin to Leiden, from Leiden to Barcelona, and have found their way into documents, papers and various laboratories.[36] However, what happens to their status if the laboratory formulates a slightly different goal? What happens when the laboratory decides to type not humans but chimpanzees – and decides to show particular interest in the differences between individual animals? Even though the markers were co-developed and tested in Lab F, and although they had been shown to work in a variety of contexts, changing the goal implies that the markers would have to prove themselves anew, in the context of chimpanzee DNA typing. The question – whether these markers are *good* genetic markers – again becomes important.

Typing chimpanzees: how far can Y-markers go?

After we had tested the first Y-chromosomal marker, I had a talk with the head of the laboratory about the results. He explained to me that the markers we were testing had never been tested before on chimpanzees, and that the alleles we had already found were so far unknown.

Continuous experiments with these markers in chimpanzees indicated indeed that none of the alleles we had found was of the same length as those found in human samples. We could conclude this now with more certainty because we ran another type of gel, an acrylamide gel using the ALF$^{®}$ sequencer. This detects the fluorescent-labeled DNA fragments via a laser beam and gives more precise identification of the allele lengths. Also, unlike the visualization on the agarose gel, the allele information is no longer physically visible but is processed via a computer and displayed on the monitor as graphs where peaks representing the alleles can be depicted.[37] It became clear to me that our interest in alleles was an interest in fragment lengths expressed by the distance between the primers. Consequently, our main objective was to compare the allele lengths found in the different samples.[38] Visualizing these alleles showed that they were not as "strong" as in humans (the bands were not as strong on agarose and the peaks were not as high on the ALF$^{®}$). My supervisor explained that this could be because of differences between human and chimpanzee DNA sequences. Additional experiments, the sequencing of the loci, confirmed this suggestion. Differences between the sequences resulted in a reduced alignment of the primers to the template DNA. Since the beginning and the ending of this region differed between humans and chimpanzees, the primers that were designed on the basis of human DNA could not attach easily. They did not yield as much PCR product and did not show very defined bands. However, although chimpanzee DNA differed from that of humans, it was similar enough to be detected by the PCR technology and to be visualized. This is not self-evident, and explains the exclamation uttered by my supervisor: "They did it. It worked!"

Once it appeared that the markers were working for chimpanzees as well as for humans, the focus of the experiments started to shift. The goal of the Dutch Primate Centre now came to the fore. We were interested not only in what the individual chimpanzees looked like for these markers, but especially in how they differed from each other. Since the difference between one allele and another is only a matter of length, our previous excitement started to wane. It became increasingly clear that for most markers the chimpanzees have roughly the same allele. The high diversity, the so-called polymorphism, reported in human individuals seemed not to be present in chimpanzees; they looked too much alike. Whereas four to seven different alleles could be detected in human populations, depending on which marker was typed, the chimpanzees showed only two alleles per marker, which was not regarded as a significant variation.

One of the seven markers tested in chimpanzees, *DYS 393*, showed four alleles and was, therefore, informative. A second marker, a curious one actually, was found interesting for another reason. Since most males have only one Y-chromosome, all marker fragments show one allele per individual. This is not the case for the *DYS 389* marker. In both humans and chimpanzees, this specific marker shows two alleles per individual, one short and one long.[39] This indicates that the primer set attaches at two different stretches of the DNA fragment. Lab F had discovered that the primer set attached to both the marker fragment as a whole and to a smaller section in that same fragment. Lab F also developed a more specific primer set for this marker to identify each allele of the marker separately, indicated as the *DYS 389* locus I and *DYS 389* locus II. Beside this peculiar feature, the variation for this marker was also not high. Like most other markers, it showed only two alleles per locus. So why was it informative? Why was it polymorphic? Another criterion for polymorphism is instructive here, namely the distribution of alleles among individuals of a specific population. Unlike the other five markers for which the chimpanzees also carried two alleles, *DYS 389* had an equal distribution of its alleles among all chimpanzees. To understand this, consider a marker for which only one individual would have allele A, while the rest of the population would have allele B; this is not as informative as a marker that shows allele A and B in a greater number of individuals. In the first case, the chance that two individuals would look alike is proportionally higher than in the second case. For this reason, *DYS 389* was found to be informative, despite the low frequency of alleles.[40] Since this marker as a whole worked for chimpanzees and since it was found informative, we tried to type the alleles of each locus separately (*DYS 389I* and *DYS 389II*). This was, however, not an easy task. In fact the primer set did not work in chimpanzees. After several attempts, my supervisor explained that the chimpanzee sequence may be quite different from that of humans, and that the primer set of the whole marker fragment may be "strong enough" to work for chimpanzees, whereas primers designed for each locus separately would not attach to the chimpanzee DNA. Consequently, chimpanzee-specific primers may have to be designed in the future to separate the alleles in each individual chimp.[41]

All together, we had one marker that proved to be informative and ready to use in chimpanzees. A second marker might become a candidate for future use, not because of a large number of alleles but because of a more even spread of the alleles. All other Y-chromosomal markers were not informative for chimpanzee DNA typing. However, as we will see below, to produce an individual profile of the chimpanzees would require more genetic markers.

Building in diversity from the start

In previous sections, I have suggested that a DNA fragment is never by itself a marker. To this end, a DNA fragment has to be aligned to a set of technical procedures. I have also argued that a target DNA fragment is not only the object of technical intervention; it has to become a technology itself. From the discussion of the chimpanzee project, it becomes clear that the goals and criteria for a particular kind of research may become involved in what a genetic marker is. So let us consider these with a view to what we can learn from them about good genetic markers.

The goal of the primate project was to identify individual primates by their genetic material. For this purpose, the laboratory sought to develop individualized data by tracing specific fragments of the DNA that, in combination, produce an individual genetic profile, the basis for a genetic passport. The central question was whether Y-chromosomal markers could contribute to the job. Do individual chimpanzees differ enough in these marker fragments to be identified? Could these markers be considered good genetic markers for typing chimpanzees? The Y-chromosomal markers are not just any kind of marker: they are already part of the laboratory's context and routines. Lab F had optimized their use in human populations, and studies of the markers for human forensic DNA typing were in progress.[42] Furthermore the laboratory as well as other geneticists had become enthusiastic about the potentials of this set of markers. Hence, its interest in how far these markers could go.[43]

Y-chromosomal markers did work for chimpanzees. It was possible to find and visualize the alleles in all individuals. Even though the visualization of chimpanzee alleles indicated a reduced alignment (allele bands were not as "strong" as in humans), the technologies and methods were powerful or universal enough to type chimpanzee DNA. However, after the first excitement, it became clear that visualization was not the main objective in this project. It was not the marker fragment as such that was informative but rather the variation in fragment length. As we have seen, at least five of these markers were no longer considered for the chimpanzee project. What does this mean in terms of good genetic markers? Lab F and the Dutch Primate Centre wanted to learn about possibilities to differentiate between individual chimpanzees, individual macaques or individual baboons on the bases of their genetic makeup. This indicates that a good genetic marker is also goal invested. In our case, only markers that worked for all chimpanzees equally and, most importantly, that revealed differences between them were found informative. One could say that only markers that came with this specific message were considered good genetic markers.[44] Such markers would have a fair chance of becoming standards in the laboratory's chimpanzee DNA typing practice. One of the markers (*DYS 389*) stood alone and was found informative whereas five others were not. It was not because this marker revealed a higher variation but because of the distribution of alleles; it was considered polymorphic, indicating that polymorphism was not only about differences but also about similarities within a given group. An allele that can be found in only one member of a group does not contribute so much to the criterion of polymorphism as an allele that can be found in 50%, for example, of its members. Thus, polymorphism is a methodologically invested category. It is made to fit the statistical means to analyze genetic similarities and differences between individuals.

From this we learn that a good genetic marker should, at the same time, contribute to the analysis of what it reveals.[45] It should produce a rate of similarity and difference according to criteria set for specific goals. Consequently, the hybrid quality of the markers discussed above, namely their being both objects of research and the technology to visualize these, is even more contingent. In a difference-producing practice, a marker is also a *methodological tool*.

The majority of the Y-chromosomal markers of Lab F turned a difference-producing practice in humans into a similarity-producing practice in chimpanzees, and were, therefore, not further considered for this project. This hints at the normative content of good genetic markers as standardized technologies. Markers enable geneticists to produce and categorize individuals, populations, humans or chimpanzees. Differences in or between these categories seem to be the outcome of the investigation. However, the analyses above made clear that, in order to study genetic diversity on the bases of genetic markers, difference has to be built into the markers from the start. Hence genetic markers that reveal variation. Given the practicalities of genetic research, studying only limited fragments of DNA and the statistical approach for analyses, *difference* is a piece of crucial information for geneticists. The chimpanzee project makes clear that what is at stake is not so much the DNA but rather good genetic markers: "standardized packages" that can do the specific jobs that geneticists encounter.[46] For there are limitations to the applicability of markers depending on the goals in question.[47] In other words, difference is not the message encoded in the DNA, but rather an effect of scientific practice. This difference-producing practice is involved in what counts as workable standard, or a good genetic marker.

To examine further the normative content of genetic markers, we will continue to follow the chimpanzee project. At this point, Lab F had decided to bring in new genetic markers. These markers were developed in other laboratories. While travelling from one laboratory to another, markers reveal the kinds of practice they embody and help to reveal the practices into which they are being introduced.

Bringing in other markers

Instead of testing other markers from the laboratory, my supervisor decided to select markers that have proved to be variable in chimpanzees in other laboratory practices. He gave me a copy of a paper and asked me to have a look at it.[48] It presented a large-scale study comparing human and non-human primates for 42 markers. The next day, we went through the paper and he explained which markers would be interesting for the primate project. Provisionally he suggested three: *FRAXA*, *DRPLA*, *SCAI*, markers of so-called disease genes.[49] These markers were promising because they showed variation in humans and chimpanzees as well as in gorillas, baboons, macaques, rhesus monkeys, orang-utangs and marmosets. Since our study was only a pilot study and given the fact that the same procedure would also have to be followed for other primates, choosing these markers would save a great deal of work in the future.

The paper gave some further information about the markers, namely the primer sequences and indications for the PCR programs.[50] The suggested primer sequences had to be ordered from a pharmaceutical company. The ordering, a very precise procedure, is usually done via electronic mail. Typing into the computer the exact sequence order of the nucleotides of the primers involves checking and double-checking the sequences. We ordered the primers and awaited them eagerly. When they arrived, we started typing the chimpanzees based on the PCR conditions indicated in the paper. None of the markers worked! I could have made a pipetting or another mistake, so we tried again, but without any success. Again no alleles. The markers worked neither for chimpanzee nor for human DNA. After having tried different samples, it became clear that we had to change the "PCR conditions", which consist of a number of variables. First the primers were considered, but these we could not change since they were the most crucial piece of information in the paper. Then we looked at the enzyme and the nucleotides, but they were standardized, supplied by a pharmaceutical company and they had worked well, as we learned from other laboratory members who had also used them. Next, the salt solution or the so-called buffer should preferably not be changed since there were too many variables in the solution itself. So the PCR program was the only possibility left. My supervisor started rewriting the programs based on the primer sequences.

A long period of trial and error started.[51] The markers came to be labeled "the experimental primers" during laboratory discussions. At a certain point programs for *DRPLA* and *SCAI* started to work – for humans, that is. Why not for chimpanzees then? Maybe the DNA we had extracted a month earlier had already started to deteriorate. It was suggested that we should test this possibility with a mitochondrial DNA (mtDNA) marker. Mitochondria have a large number of small, circular DNA molecules and their alleles are much easier to detect. The mitochondrial marker showed very strong alleles. So it was not the quality of the DNA. Again new PCR programs were suggested. I asked my supervisor: "Since the programs worked for humans, why not keep them and change the salt solution instead?" Again he objected because he wanted to keep that standardized.[52]

At the next laboratory meeting, we reported on the problems we had encountered. One laboratory member suggested contacting the authors of the paper and asking them about their experiences: "They might have other laboratory conditions." A discussion about laboratory conditions ensued. Another member, who was also working in a diagnostic laboratory, reported that they used different salt solutions for different markers (whereas our laboratory had a standard salt solution) and that they used standard PCR programs (whereas we had marker-specific PCR programs). During the meeting, I asked about this difference. Laboratory practice was the key here. In a diagnostic laboratory, samples of individuals are kept strictly separate. Individuals are usually screened for a number of markers, and experiments are conducted on one individual at a time. The most efficient way to do this would be to run one PCR for all the markers at the same time. The variable in this kind of laboratory would be the salt solution. Our laboratory studied individuals and populations and compared these for one marker at a time.[53] So an efficient PCR run consisted of as many samples as possible from different individuals to be typed for one marker. Therefore, our laboratory had different PCR programs for different markers. After this discussion, I understood the laboratory's general motivation and I stopped asking questions about salt solutions.

More PCR programs were designed and tested. Making the markers work became an obsession. One laboratory member suggested that we conduct a search on the Genome DataBase (GDB) and look for other markers with more marker information. This was no option. They had been shown to work in the scientific paper so they should work here as well, we reasoned.[54] After a while, we started to have some results in *DRPLA* and *SCAI*, and as expected they proved to be variable for the chimpanzees: six alleles were found for *DRPLA* and five for *SCAI*. For the third marker, *FRAXA*, my supervisor contacted colleagues at the neighboring diagnostic laboratory, who had a great deal of experience with this marker. However, it turned out that their primer sets were radioactively labeled, whereas we were working with fluorescent groups. Consequently, it did not make sense to use their protocols. Yet another laboratory in the Netherlands was contacted and they sent us different PCR programs, but nevertheless advised us to drop this marker because it was too hard to type. It appeared that the problems we had with this marker resulted from particularities of the DNA fragment. The intricacies of the fragment prohibited its amplification.[55] Specific nucleotide repeats in the sequence caused the fragment to fold in complex ways, which made it difficult for the standardized PCR technologies to denature and copy the fragment.[56] Their protocols suggested a special type of nucleotide to expedite the amplification. With some modifications in the PCR program suggested and the addition of these special nucleotides, we started to have results with this marker. Finally, we were also able to detect four alleles in the chimpanzees.

Altogether, we then had a set of five and perhaps six markers that were ready to use for chimpanzee profile typing and to be tested on the rest of the primates.[57] This set proved workable for chimpanzees and it met the criteria for profile typing because it showed a considerable variability in each marker fragment.

In the meantime, the highly organized laboratory, with clear-cut procedures and protocols, had changed dramatically. PCR machines as well as the ALF® sequencer would be overbooked; colleagues would wish us good luck with breeding chimpanzees. Others received presents of the (unknown) trademark Monkey Jewellery Inc., and the previously well-organized and well-marked set of PCR programs would include programs such as Sky (*SCAI*), Touch down (*SCAI*), amade1 (*DRPLA*) and Hot PCR (*FRAXA*).

Markers: mediated practices

Just like the Y-chromosomal markers, the newly introduced markers have found
their way into scientific papers and eventually into different laboratory practices. However, once introduced to Lab F, they faced problems. The experiments
that followed made clear that markers are more than just a DNA fragment or
a variation that could be found in different individuals; they also showed that
the constituents of markers that were addressed before were not self-evident,
"universal" or problem free. The last had not become clear while testing the
Y-chromosomal markers because they were already operative in Lab F.[58] The
technologies and practices developed with the newly introduced markers proved
to have decisive consequences for their visualization and their applicability for
the chimpanzee project. The process of making them work in a new context
displayed the various socio-technical components implicated in markers. The
markers were dependent on a network where knowledge, experience and technical details could be exchanged, and where specific additives could be purchased.
Taking this network into account, it could be said that what makes a marker a
marker could as well be a PCR program, a salt solution, a radioactive-labeled
primer, a fluorescent-labeled primer, a complexity in the DNA fragment or a
specific type of synthesized nucleotide in the reagent.[59] All these aspects may
become crucial and dominant when markers start to move from one socio-
technical context to the other.

A marker is thus dependent on the kind of world that *can* be introduced to
a laboratory and the kind of alignment that *can* be established there. Accom-
modating the visualization of the newly introduced markers was dependent
on similarity between scientific practices. It required that the practices of the
laboratories where these markers came from were similar enough to those of
Lab F.[60] However, as we have seen, making the markers work also required a
reordering of the practices they embodied. The stable form of their socio-
technical components, namely the protocol described in the scientific paper,
had to be loosened in order to make space for modification and adaptation.[61]
In addition, the laboratory's ordinary way of doing things had to be opened up
and was shown to consist of different procedures, technologies and types of
research, which to a certain extend could be rearranged as to make the markers
work. As we have seen, however, neither the flexibility of the markers nor that of
Lab F were infinite. Some aspects of the protocol had to remain intact, for example the primers suggested in the paper or the application of PCR technology.
Lab F appeared to be more inclined to change its PCR programs and reluctant
to change the buffer (the salt solution). The more rigid or immutable features of
the daily work were not only articulated by those who conduct the experiments
but were also embedded in the organization of that work. Immutability was

delegated and built into standard procedures and the handling of, for example, the PCR machine and the PCR reagent. One could say that, given the traffic in objects and technologies between laboratories and taking into account the procedures of making things work described here, a marker might be seen as an enactment of *local mediations of scientific practice*.[62] Consequently, it is in and through such local mediations in different places and times that a marker may become a standard, a *good* genetic marker.

Standards bear the promise of making practices comparable over space and time. However, given the hard work involved in making standards operative in local settings, technologies may not succeed in becoming implemented. For example, the markers described above risked, at least at one point, being replaced by others. However, not only did they have a long history in the field of medicines and genetic research, they had also become part of the global discourse of diversity studies on primates of various kinds. There may well have been hundreds of possibilities for typing chimpanzee DNA, but the stable form in which all these aforementioned qualities were presented, namely in a key scientific publication, made them and not others the markers of choice. Here rests the power of standards. It is, of course, in many cases easier, less time and money consuming, to introduce standardized technologies: technologies that have proved themselves elsewhere and which are broadly applied by peers.[63] This enlarges not only the rate of success in applying them but also the chance that the scientific data produced through them can travel farther: even though scientists do not know beforehand how far their data will go, or how far they would want them to go. Yet there is more to the power of standards: namely their sheer "presence".[64] This is their power to "replicate", to "order" social worlds[65] and to render specific versions of the world into the most obvious ones. They simply make themselves self-evident and so close to hand at the cost of many other possibilities. For example, in the field of genetics, and also for the purpose of diversity studies, different technologies may be at hand, but their actual use is grounded in a tension between more- or less-stable practices, and more- or less-loose ways of getting daily laboratory work done. It is a tension through which some technologies become privileged over others and certain ways of knowing genetic diversity become more relentless than other ones.

Returning to the round-table discussion

In fact the round-table discussion introduced at the beginning of this chapter illustrates the organized and negotiated character of scientific work and the traffic in markers. Let us consider the remarks quoted above and examine these in the light of what we have learned from the chimpanzee project.[66]

A list of markers should be made as an indication to newcomers in the field and in order to compare the different data

Obviously geneticists do not work alone but within a field. Even more, this field is populated not only by people but also by technology. Comparability of data and applying similar genetic markers are, like the geneticists themselves, crucial to the field and contribute to its existence.

We need to have some consent about the markers so as to compare the results

Emphasizing consent about which markers should be on the list suggests that knowledge, although a product of local practices, becomes meaningful only within communities. This indicates that local scientific practices are informed by criteria that transcend local contexts. The global is thus enacted in the local. Since the comparability of data is not self-evident, it should be built in at an early stage of experiments. This quality is delegated to markers. Hence the "global" aim of scientists, namely that of working together, is to be implemented in local practices to produce this possibility beforehand. Standardization via a list of markers should do the job. But what about content? Does not a list of markers predefine what type of knowledge will gain pride of place?

Criteria for markers should be that they show variation between populations

From the chimpanzee project, it had become clear that markers are actively involved in what genetic diversity is. Markers are hybrids. They are technologies of visualization, methods of analysis, as well as objects of research. Therefore, similarities and differences between one individual and the other, or between one population and the other, are marker dependent. In fact, markers help to establish the boundary between such categories. This was a key argument of Chapter 2. The standardization of markers for the purpose of diversity studies has as its effect a standardized approach to what individuals and populations are. Standards thus establish and reify such categories.

We need markers that are selectively neutral to different population structures

The suggestion made here is that markers should work for all populations in the same way and that they should provide information without respect to the populations studied. This means that the markers of choice are inherited independently and are not submitted to within-population structures, which are the results of biological or social constraints. This presupposition is called *random mating*. It underlines that geneticists do not study genetic lineage as such (i.e. how individuals and populations are related), but rather a version of lineage that is theoretically invested by genetic markers.

Preferably markers that do not require the use of radioactivity

Choosing certain markers and not others for safety reasons, such as not using radioactively labeled primers, has an impact on the type of knowledge that is possible. The chimpanzee project is instructive here. Lab F could not learn from the practice of a neighboring laboratory because they used radioactively labeled primers. So neither the comparabilities of data nor safety measures are external to knowledge. They too determine which technologies become most favored in learning about genetic lineage.

What we need right now is a list of priority markers

Prioritizing some markers emphasizes the variety in local practices and the need to attune these practices in order to work together and to profit from knowledge and expertise developed elsewhere in the field. At the same time, a list of priority markers indicates that there are tensions involved in doing so.

What is the use of such a list if people aren't working with it anyway?

Whose markers will be on that list is crucial. A list will not only have consequences for the type of knowledge but also for the type of practice transported from one laboratory to the other. As we have seen above, laboratory work tends to evolve according to a "practical conservatism".[67] Lab F, for example, started out testing its Y-chromosomal markers first, because these were already part of its routines. In addition, even the newly introduced markers had to be plastic enough to be modified according to the laboratory's way of working. Such routines consist not just of standardized procedures but also of social technical networks, where there is an exchange of technical information and knowledge. Therefore, introducing new markers might well mean introducing the need for socio-technical networks. Equally, what if a laboratory is working successfully with radioactively labeled primers?

One could recommend things now, but it would be preferable to choose for a democratic procedure, such as people reacting from their own experience of research

Procedures of standardization are never neutral. The suggestion made in this comment, namely to go from the local to the global, recognizes the diversity in the organization of scientific work and aims at exploring possibilities for standardization "from below". It acknowledges the politics involved in standards and the difficulties of changing existing practices. Above I have addressed "presence" of standards and the power involved in that. For example, the Y-chromosomal markers became standards within two years and the number of publications based on these markers is still growing. All without any organization or group of scientists ordering the world to use them. However, the success of Y-markers might be ascribed to the various different goals that can be reached by them, such as paternity testing, forensic science, genetic lineage and diversity. Even when most of the time, once they are being applied somewhere, standards tend to go unnoticed, they have the tendency and power to replicate, to enlarge their field of use, enrol more and more users and other actors and, consequently, to affect local practices.

What are good genetic markers?

Markers are not merely local nor are they entirely global. They are rather the effect of mediations between them. Reflecting on the round-table discussion in the light of the chimpanzee project has underlined the normative character of such mediations. Introducing good genetic markers, markers that could go unnoticed under the umbrella of a priority list, is bound to do more than establishing collaborations between geneticists. For, even if the path of scientific work were to be smooth, standardization is not neutral to knowledge. Introducing standardized technologies co-determines what can be known about genetic diversity and helps to stabilize particular objects of study, whether this is the individual or the population.

Conclusions

Genetic lineage and diversity are dependent on markers. The concern of this chapter was to investigate what a genetic marker is. In this investigation, the definition of a marker pointed in the direction of the DNA. Laboratory practice, however, suggested other aspects for learning about markers. We learned that markers are enacted as a variety of things, such as protocols, PCR programs and chemical solutions; aligning these in more or less stable ways may turn a marker into a good genetic marker. A marker can thus be termed a *local mediation of scientific practices*, through which humans, technical devices, chemicals, DNA and procedures to handle it are linked in a specific way to produce the marker. Markers and their use may well be standardized. Yet making them work in a specific laboratory requires versatility in protocols and reagents, both similarity to and flexibility in the practices to which the markers are introduced, as well as local networks through which knowledge and expertise can be exchanged.

Standardized or good markers need to "incorporate and extend" such networks and routines.[68] Consequently, enacting markers in a new context is dependent on the kind of world that *can* be introduced to a laboratory and the kind of alignments that *can* be established there.

This has implications for studies of diversity and lineage. This very quality of markers not only puts constraints on what *can* be standardized and how, but also on our ways of learning about genetic diversity. To do this, various technologies may be at hand, but making things work in a variety of practices equally evokes questions about which technologies will have pride of place in learning about diversity and lineage. Rather than a universal tool, a *good* genetic marker is a highly invested category in which genetic diversity resides.

Notes

1. For a similar narrative strategy and analysis, see Mol and Law (1994).
2. These as well as other contributions to the discussion were elaborated in a document that was intended to function as a guideline on the choice of markers (see note 38 in Chapter 2). In the end, the document was not issued but circulated, via email correspondence, among a number of geneticists who contributed to the document and commented on it. Bertranpetit, the editor of the document, presented the results, in a way a list of priority markers, at a meeting of the *European Human Genome Diversity Project: Regional Committee* in January 1996 in London (Bertranpetit, 1996). My references to this document are based on the copy that was distributed by Bertanpetit via email on 3rd February 1996.
3. Under the heading *Presentation*, Bertrantpetit (1995: 1) stated: "In the program, a large amount of time has been reserved for discussion as consensus will be sought on the technical advantages and informativeness of the various marker technologies. It is our aim that presentations and discussion will mainly focus on DNA markers."
4. During *The International Planning Workshop* held in September 1993 in Porto Conte, Sardinia (HUGO 1993), the fine-tuning was on the side of population, such as the sampling procedures (criteria for sampling, how many individuals per population and how to inform the individuals about the goals of the scientists) and which populations should be studied first. The endeavor then was to produce a "priority list" for the populations (HUGO, 1993: 12–33). The issue of population and the sample strategy of the Diversity Project was the topic of a workshop held in October 1992 at Pennsylvania State University. Furthermore, in his personal communication, Bertranpetit argued: "Other crucial issues in genome variation analysis are not considered here, comprising the choice of populations, sampling strategy, availability of samples through central repositories (of DNA but plasma may be useful), fingerprinting of reference specimens, data bank results, numerical analysis of results and many others."
5. See also Chapter 6. For an overview of laboratory ethnographies, see Knorr-Cetina (1995).
6. For an excellent study on routine-like technologies, see Jordan and Lynch (1992). I apply the notion of enactment to make clear that 'objects' do not exit by themselves

but are enabled by humans, materials and techniques. What these objects are made to be, which version of them is being performed, is, therefore, dependent on practices. Moreover, this suggests that objects are not stable as such, and that their 'existence' is dependent on the successful work of humans and non-humans; see, for examples, Hirschauer and Mol (1995); Law (2000); Mol (2000).

7. See, for example, Latour (1987).
8. Latour (1990) speaks in this respect of "immutable mobiles." On standards in science and how they enable coordination between different scientific groups, see, among others, Fujimura (1987, 1992); Timmermans and Berg (1997).
9. Bowker and Star (1999: 34) termed such events "infrastructural inversion."
10. From the *Glossary* in Kevles and Hood (1992a: 381). A similar definition can be found in the *Glossary of Terms* in Department of Energy (1992: 36), *DOE Human Genome Project: Primer on Molecular Genetics.*
11. In molecular biology, a "DNA marker" was originally a synthesized DNA fragment of known size, through which the molecular weight of target fragments could be determined, see Hartl (1995: 379).
12. Anon (2001a,b).
13. The complete human genome, coding and non-coding DNA, consists of 3500 million base-pairs; the percentage of coding DNA may differ depending on which literature is considered. The figure of 5% is based on Kevles (1992: 24). Recently, we have learned that the number of genes is even less than most geneticists expected.
14. The significance of this difference can be seen in the criticism voiced about the (physical) mapping of the human genome. The critiques considered it too time consuming and costly and not informative as such; the so-called large-scale sequencing without a directed interest in specific genes was seen as a waste of effort. See, for example, Kevles and Hood (1992b).
15. See also Rabinow (1996a: 1–5).
16. In the laboratory, agarose gel electrophoresis is the first and most practical method of visualizing the DNA fragment; it takes less time than using an automated sequencer and is less expensive. The detection of alleles with an automated sequencer is also based on electrophoresis; instead of ultraviolet it requires laser light for visualization. When an electrical current field is applied to the DNA fragment, which has a plus and a minus pole, it will "migrate" over the gel under the electromagnetic field created. The time a fragment needs to migrate is related to the "molecular weight," the length of such a fragment. Typically the visualized alleles on agarose gels are referred to as bands and in a sequencer they are referred to as peaks, indicating what can be seen or – better – how the alleles are represented. The automated laser fluorescent sequencer (ALF[®], Pharmacia) produces not only the raw data but it also processes the data with an inbuilt computer. One could say that with this feature the ALF[®] fuses what Amann and Knorr-Cetina (1990) have "identified" as "data" and "evidence," which are defined by the different practices they involve; the ALF[®] sequencer produces both data and evidence simultaneously.
17. These ladders, also called sizers, serve as molecular weight "markers"; see Hartl (1995: 379).
18. See Amann and Knorr-Cetina (1990: 92) for an elaboration on gatherings of laboratory members around visual objects, depicting what is visualized in the process of talking about this. Emphasizing the interconnectedness between the laboratory talk and the (visual) object of research, Amann and Knorr-Cetina refer to "the talk" as "the machinery of seeing". See also Hacking (1992).

19. Analogously, Annemarie Mol (2000) argue that the "thickening of the intima" in cases of "atherosclerosis is not all alone" either but is dependent on visualizations, chemicals and tools added to the tissue to be analyzed as well as the skilled eye of a technician. My use of alignment is akin to that of Joan Fujimura's. Whereas Fujimura (1987) takes the broader context of laboratory work and social interactions into account, I focus more on technologies and their practicalities in this context.

20. Lynch (1990).

21. Similarly Hans-Jörg Rheinberger (1997: 275) argued that the tools of molecular biology are involved in the object of research, so that they become macromolecules themselves: "Die Scheren und die Nadeln, mit denen Gene geschnitten und gespleißt werden und die Träger, mit denen man sie transportiert, sind selbst Makromoleküle."

22. See Lynch (1990) and Rheinberger (2001).

23. This is the former TNO-Dutch Primate Centre. TNO is the Dutch Organization for Applied Scientific Research. In the mid-1990s, the Dutch Primate Centre became a semi-private organization aiming at becoming the world's largest primate center. This endeavor may guarantee considerable financing from the Dutch government.

24. Furthermore the head of the laboratory used to hold a position in the Leiden TNO, where he conducted and guided research into heart disease.

25. The Y-chromosome is one of the so-called sex chromosomes and is normally inherited by men only. Daniel Hartl (1995: 445) defined polymorphism as follows: "The presence in a population of two or more relatively *common* forms of a gene, chromosome, or genetically determined trait" (emphasis added).

26. In June that year, when we had a meeting with the chairman of the Primate Centre, he seemed to be somewhat disappointed that we had started with the chimpanzees. He would have liked us to start with the macaques instead, for they are the most frequently used in laboratory research and thus the most interesting object of trade.

27. This was the Institut für Gerichtliche Medizin at the Humboldt University, Berlin.

28. The Y-chromosome has not been as studied by population geneticists, but interest in this chromosome has been growing. In the small number of publications during the early 1990s, it was argued that the Y-chromosome did not show considerable polymorphism (genetic variation). It was only towards the middle of that decade that some work provided evidence for polymorphisms on the Y-chromosome. At the 1995 conference *Human Genome Variation in Europe: DNA Markers* in Barcelona, a group of young population geneticists (Mark Jobling, Peter de Knijff, Chris Tyler-Smith and Antti Sajantila) working on Y-chromosome variation were highly appreciated for their "pioneer work" on the Y-chromosome. They all participated in the panel for the conference's round table discussion and were all invited to participate as co-writers of the report about genetic markers, Bertranpetit (email personal communication).

29. Roewer *et al.* (1996: 1031). An allele should here be understood as a term expressing different sequence lengths found for one marker in different individuals and measured by the distance between the primers (see also note 31). A microsatellite is an alternative name for a short tandem repeats (STR) in a marker region. For example, the nucleotide sequence ACTACTACT would be a tandem repeat consisting of three nucleotides ACT. The tandem may vary between two and five nucleotides, see Watkins, *et al.* (1995).

30. See Chapter 2 for an elaboration of the interdependence of markers and population.

31. Roewer *et al.* (1996: 1032). Haplotypes, as referred to in this paper, are conceptual genetic profiles (a combination of alleles of different loci for one individual) used as a statistically informative measure to compare population similarities and differences.
32. de Knijff *et al.* (1995).
33. Section 8 of the emailed personal communication by Jaume Bertranpetit addressed Y-Chromosomal markers based on a collaboration between P. de Knijff (Leiden) and M. Jobling (Lancaster).
34. de Knijff *et al.* (1997); Kayser *et al.* (1997). These papers are the result of the first meetings of the *Y-Users Workshop* held in Berlin in 1996, which was initiated after the 1995 Barcelona conference *Human Genome Variation in Europe: DNA Markers*.
35. See Latour (1990).
36. One could say that these markers have become what Bruno Latour (1990) has called, "immutable mobiles" in the sense that they produce and are products of a network of scientists and laboratories, technologies and tissues, literature and methods. For the limitations of "immutable mobiles," see Mol and Law (1994); for a critique, see Fujimura (1992).
37. Reading the information produced on the ALF$^{®}$ definitely requires training. An unskilled viewer may see a broad range of peaks in complex graphs, where a trained eye would locate the actual allele peaks within a couple of seconds. Also, the ALF$^{®}$ comes with a software program, the ALF manager, which allows practitioners to determine the allele length and to "polish" the graphic representations by zooming in on the target peaks and amplifying them on monitor and printouts. In accord with Michael Lynch's (1990: 161ff.) treatment of scientific visual object (specifically of photographic images and diagrams), one could term this conduct an "upgrading" of the visual object genetic marker.
38. Hence, for Lab F, the difference between the one allele and the other is a difference in length. However, an allele may mean different things in different contexts. For evolutionary studies, an allele may refer to the sequence proper, and differences between alleles may be categorized according to differences in nucleotides (e.g. by mutations in the sequence). Hence not the length but the sequence order is relevant here. In disease studies, the genetic code, namely the triplet of nucleotides coding for an amino acid or a protein, is more relevant in determining alleles. From this perspective, the difference between one allele and the other may be a difference in triplets that code for the same acid or protein (e.g. both GAA and GAG code for the amino acid glutamine). Furthermore, a frequently used definition of an allele is "One of several alternative forms of a gene occupying a given locus on the chromosome. A single allele for each locus is inherited separately from each parent, so every individual has two alleles for each gene" (Kevles and Hood, 1992a: 375). Note that this definition contains a bias. First of all, alleles on the Y-chromosome are inherited from the father only, so the individual does not have two. Second, there is a bias for coding DNA in this definition: the definition refers to genes, so the largest amount of DNA, the so-called "junk DNA," is not included here. Third, there is a bias for nuclear DNA: chromosomal DNA is exclusively nuclear DNA. Mitochondrial DNA, which is outside the nucleus on the mitochondria, is excluded.
39. Since males usually have only one Y-chromosome, it is strange to find two alleles for a marker in an individual. For this marker, it appears that the primer set (the synthesized sequence indicating the beginning and the end of the marker

fragment) matches two loci on this chromosome and, therefore, shows two alleles per individual.

40. This is because profile typing is based on a rate of difference rather than on absolute differences, also called matching likelihood estimations. Combined with statistical analysis, a rate of difference will allow the laboratory to identify individual chimpanzees, see Chapter 2 for the individualization of genetic profiles and matching likelihood estimates.

41. Provisionally, for the purpose of sequencing, we diluted the two allele fragments from the agarose gel. After electrophoresis, the specific allele fragments were "physically" locatable on different parts of the gel; by cutting the gel into pieces, we could retrieve the fragments separately for further experiments.

42. See Roewer *et al.* (1996: 1032).

43. Lab F was at that time (1996) in the process of having more and more of these Y-chromosomal markers accredited by the Dutch Board of Accreditation. This would allow the laboratory to apply these markers for all forensic DNA cases.

44. It is interesting to note that are there are two categories of markers, referred to as mini- and microsatellites, emphasizing the transmission of information involved. The types of marker discussed in this chapter are also called microsatellites.

45. This is comparable to Lynch's (1990: 163–168) argument about diagrams. Diagrams not only represent what can be seen but also what can be claimed about a biochemical structure.

46. See Fujimura (1987, 1992, 1995).

47. Also Roewer *et al.* (1996: 1032) indicated that Y-chromosomal markers may work for certain forensic tasks but not for others. Even though the variation found in humans is high, it was not possible to identify all individuals studied on the basis of these markers only. Therefore, markers can lose their usefulness when the goals change.

48. Rubinzstein *et al.* (1995).

49. Disease genes are studied extensively, obviously, because they are correlated to a specific disease. The abbreviations stand for the diseases they are associated with. *FRAXA* stands for a type of the fragile X syndrome, *SCAI* is associated with spinocerebellar ataxia type 1 and *DRPLA* stands for dentatorubral-pallidoluysian atrophy (Watkins *et al.*, 1995: 1485).

50. The paper indicated only the so-called annealing temperature. On this basis, the PCR programs could be deduced. The PCR programs are based on different numbers of cycles expressed in the time and temperature of denaturing and annealing. Primers attach at different temperatures to the DNA fragment according to their sequences. These annealing temperatures have to be optimal for an adequate copying of the DNA. "Denaturing" of the DNA means that the double-stranded and helix-shaped DNA is pulled apart into two single straight strands. This optimizes the possibility for the primers to attach at the beginning and the end of the target sequence and thus the copying of the DNA. "Annealing" is the movement of two single strands of DNA to form one double-stranded DNA. Just before doing so, the primers attach to the single strands, preventing them from clinging to one another and forcing the enzyme polymerase to use the "free floating" nucleotides in the solution in order to make single strands into double ones again.

51. This indicates that genetic markers cannot entirely be understood as immutable mobiles, as I suggested earlier (note 36); once moved to another context, they prove to be *mutable*. See also Mol and Law (1994).

52. With Jordan and Lynch (1992: 93–94), one could term this a kind of "practical conservatism." If things have proved to work in previous experiments, laboratory members tend to be "superstitious" about changing their mode of working; how the number of variables work together cannot be understood in rational ways only.

53. In Chapter 2, I have shown that Lab F cannot "know" an individual without a concept of population. Thus individual genetic makeup and that of a population are mutually dependent.

54. Although the forensic work of Lab F can be categorized as routine work rather than fundamental science, the experiments I describe here echo Karin Knorr-Cetina's claim (1981) that laboratory action is about making things work. Moreover, the argument that the markers introduced to Lab F should work since they have been shown to work in another, presumably similar, context underlines that scientific work may rely on a variety of previous practices as long as a kind of similarity can be presupposed between these practices, even at the risk of failure and the loss of time and resources (Knorr-Cetina, 1981: 57–63). For a variety of stories about making a routine-like technology work (such as the plasmid preparation, a cloning technology), see Jordan and Lynch (1992).

55. The so-called secondary structure of the DNA makes it difficult for the fragment to denature. Therefore, the whole PCR procedure was inhibited, the marker fragment could not be copied and consequently it was not possible to detect the alleles.

56. *FRAXA* contains a CGG repeat. The nucleotides C and G also complement each other in the fragment, allowing hydrogen bonds (relatively easily broken) to form between the two strands, giving the base-pair C–G. The large number of C–G bonds in this fragment introduces many such linkages and these impose a structure, known as the secondary structure, to the DNA. The double-helix structure of DNA occurs when two single strands link by hydrogen bonding between base-pairs to form two chains coiled around a single axis. This secondary structure makes it harder for the PCR method to work as it must first denature (straighten and pull apart) the DNA fragment.

57. This set consists of *FRAXA*, *DRPLA*, *SCAI*, one mitochondrial DNA marker, *DYS 393* and perhaps *DYS 389I* and *389II*.

58. For how concepts as well as laboratory practice are involved in genetic objects such as "probes," see for example Fujimura (1992); Jordan and Lynch (1992). Moreover, Fujimura (1987) argued that standardization increased "the doability" of laboratory experiments and that it conversely reduced laboratory uncertainties and time investments. In a way, the newly introduced primers obliged Lab F to open its "standardized packages" of doing PCR.

59. See Rheinberger (1997); Mol (2000).

60. See, Latour (1988); Timmermans and Berg (1997); Bowker and Star (1999).

61. On how loose networks can be turned into stable ones and on how laboratories accommodate both stability and instability in medical interventions: see Singleton (1998). On how objects that are manifold can be coordinated and enacted as singular, see Law (2002); Mol (2002).

62. This notion is akin to Michael Lynch's 'locally achieved agreement' (1985: 155–170).

63. See note 58; Bowker and Star (1999); Star (1991).

64. Star (1991: 38–39) argued that the power of standards is not only about what is included or excluded, which actions are made possible and which not, but also

about the very *fact* of standards. Their presence in the landscape and the various kinds of relation that they help to establish through their presence.

65. See Law (1994).
66. For the sake of this argument, I have changed the order of the statements.
67. See Jordan and Lynch (1992).
68. Timmermans and Berg (1997).

4

Naturalization of a reference sequence: Anderson or the mitochondrial Eve of modern genetics

Introducing the argument

Technologies are locally mediated scientific practices. And standardized technologies are not fixed but "achieved in process." This was the topic of Chapter 3. Often, however, technology is treated as an autonomous entity; once produced it leads a context-independent life and is rendered self-evident in various different contexts. It becomes naturalized. The naturalization of technology not only presents it as a clearly defined entity; it also obscures the normative content of technology. Questions about how a technology might affect the world, thereby move to the background. This chapter deals with the naturalization of a standardized technology. It examines both its ideological content and how it is made into an autonomous entity. The argument is that since technology is always produced *somewhere* and is made to work in a *specific* context, its naturalization also requires work. What kind of work this is and how it is carried out is the main concern of the present chapter.

In February 2001, the sequence of the complete human genome was presented to the world.[1] We also learned that the work of the Human Genome Project was not yet complete and that substantial parts of that genome were still being mapped. However, in 1981, a complete sequence of another human genome was produced in Cambridge. Not a sequence of nuclear DNA, but that of the mitochondrial DNA (mtDNA). Ever since its publication, this sequence had become the reference genome for geneticists. It is a standardized technology that is known as *Anderson*, referring to the first author of the scientific paper. It (nowadays) can be downloaded from the Internet and appears on the computer screen as a text. Anderson plays a central role in studies of genetic diversity.

The argument of this chapter is organized around this *reference sequence*. I will first address the kind of work it enables in laboratories. Then I will investigate how it was produced and the kind of practices, technologies and cell material that were required for that. In doing so, I will pay special attention to issues of race and sex differences in the context of genetic diversity. Finally, I will argue that the naturalization of this reference genome and its functioning as an unproblematic standard involve the universal theory of mitochondrial inheritance to which it is connected.

Neanderthal: the sequence

It is November 14 1996 and we are at the Laboratory for Evolution and Human Genetics in Munich, which I will refer to as Lab P (after its head Pääbo). This morning I entered the small laboratory, where I have my bench. I had put my coat and bag in the locker down the hall, had a quick look at my laboratory journal to recall the program for the day and went to get myself a cup of coffee. As usual, I walked through the small writing room, situated between the small laboratory and "the tea room." I noticed champagne glasses, empty bottles, cigarette boxes and full ashtrays all scattered about the long writing desk. The cigarettes especially drew my attention. People may have dinner in the laboratory or sleep in the laboratory, but smoking there? That just never happens. A colleague came in and asked me whether I had heard about Matthias Krings. There had been a party last night she told me, because he had sequenced the Neanderthal for the second time and found the same DNA sequence.[2] I became aware of Krings' absence. Normally he would be working day and night. My colleague went on telling me about the excitement of the event and described how he and the archeologists, who had made the Neanderthal bones available, were looking attentively at the computer screen while the gel was run on the ALF®.[3] As it became visible, the sequence turned out to be the same as the previous one Krings had found. Thus it proved that what they had found could be called the first ever mtDNA sequence of a Neanderthal. They decided to have a small party and invited other people who were working late to join them.

Krings came in only the next day. I went over to congratulate him. We started talking about the Neanderthal. He told me that the bones had been found in the west of Germany (Düsseldorf) back in the middle of the nineteenth century and had been supplied to him by archeologists from the Rheinisches Landesmuseum (Bonn). Talking about the forthcoming article, I asked him whether he would publish his results in a specialized journal in the field of archeology. He made it clear that *Nature* would be a more appropriate place for it[4] and told me that further (statistical) analyses would have to be conducted to learn more about how the Neanderthal relates to humans. Before publication, however, the experiments had to be repeated in another laboratory to ensure that the sequence was not a "local" artefact.

Once the Neanderthal DNA had been sent off to a laboratory in Berkeley, California, a period of trial and error started, until finally a sequence was acquired that was similar to the one produced in Lab P.

The significance of the first sequence

Sequencing DNA – in other words, unravelling the sequence order of the DNA building blocks – would involve many uncertainties and guesswork if it were not assisted by already existing sequences. As indicated above, the topic of this chapter is a reference sequence, a sequence that does such a job in the field of population genetics. The case of the Neanderthal is an example of such a reference. It shows the work involved in producing a first sequence: the hard work of Krings, who had labored day and night to sequence it; the availability of the samples, for which the suppliers have travelled many times from Bonn to Munich; and the celebrations indicating the role it may play in genetics, not only because it is a Neanderthal sequence but also because it is its very first sequence. The question I raised concerning publication of the Neanderthal results was based on the idea that this new insight may be of special interest to those working in the field of archeology and paleontology. Krings' answer, however, made clear that it had a much wider relevance for geneticists interested in genetic lineage. It hinted at the technological achievement involved in the sequencing.[5] The sequence had the potential of becoming a piece of technology, a reference for other possible sequences to come. The relevance of such a sequence was already indicated in the work of producing it. Since the Neanderthal had no reference sequence, a standard to compare it with, Krings had to acquire the sequence twice, and in Berkeley the whole procedure had to be repeated. All this work brings to the surface the absence of a fixed standard by which the Neanderthal sequence could be confirmed. The sequence was instantly treated as more than a Neanderthal sequence. It became a tool of comparison between modern-day humans and Neanderthals; a tool through which genetic proximity and distance could be addressed. These treatments are indicative of the reference sequence and of what it enables in laboratory work.

Just like the mtDNA sequence of the Neanderthal, that of modern-day humans was produced somewhere and embodies the practices that helped to produce it. Yet it has become a standardized tool, a text, available to laboratories all over the world. In this chapter, I will address how the reference has escaped the "burden" of locality to become a naturalized tool. To do so, I will move back and forth through time, from a contemporary laboratory practice to more "historical" accounts of how DNA work was done in the early 1980s, and then back to the 1990s to view how theories of mtDNA inheritance are involved in the work of this reference. Before doing so, let us have a closer look at mtDNA and its particular role in studies of populations and genetic lineage.

Not in the nucleus: mitochondrial DNA

In April 1953, James Watson and Francis Crick published the molecular structure of DNA in the journal *Nature*. The abbreviation DNA stands for *deoxyribonucleic acid* and indicates not only the chemical and physical structure of DNA but also its location in the cell, namely in the nucleus.[6] This is not entirely true for mtDNA. Even though it consists basically of the same building blocks as nuclear DNA,[7] mtDNA is located in the cytoplasm rather than the nucleus. More precisely, it is situated in the mitochondria, organelles that provide the cell with energy. Moreover, unlike the nuclear DNA molecule, which is a large (3.5 billion base-pairs)[8] and stretched out over 46 chromosomes, mtDNA is small (16 500 base-pair) and circular. Each human cell contains one nucleus but may contain thousands of mitochondria. Consequently, compared with the two copies of nuclear DNA (one stemming from the father and one from the mother), each cell may equally carry thousands of copies of mtDNA.[9]

The location of mtDNA in the cytoplasm creates another major advantage for studies of genetic lineage. Why? The answer has to do with reproduction and genetic inheritance. All mitochondria passed on to offspring come from the mother. The sperm consists predominantly of a nucleus and it has only a small number of mitochondria (about 50 copies) in the mid-piece; these mitochondria do not enter the egg cell during fertilization. Therefore, males inherit mitochondria solely from their mothers and cannot pass these on to their offspring. Only females do so. This means that geneticists can trace mtDNA inheritance through maternal lineage. Whereas nuclear DNA recombines during meiosis, when a great part of the genetic material coming from both parents is being reshuffled, mtDNA does not. It comes from one individual only. This single parental inheritance allows for genetic studies that look way back into history.[10]

Not only the patterns of inheritance but also the size and the number of mtDNA molecules are seen as advantageous in genetic research. Whereas the sequence of the human nuclear genome has just been compiled in the HGP,[11] the complete human mitochondrial genome had been sequenced as early as 1981, by one laboratory. Additionally, because of the number of molecules per cell, it is possible to study the mtDNA of extinct as well as living species: hence the Neanderthal sequence. Bones or fossils may still contain some intact mtDNA since there are thousands of copies available, whereas the nuclear DNA, consisting of only two copies, is difficult to retrieve and is usually destroyed with the passage of time. This feature of mtDNA, namely that it can be studied in different species and permits comparison of species from an evolutionary perspective is, in a way, captured in the name of Lab P: The Laboratory for

Evolution and Human Genetics. It accounts for the variety in research projects that can be found in this laboratory.

Mitochondrial DNA in the laboratory

Between October 1996 and April 1997, I was in Lab P to conduct a participant observation on human genetic diversity. After that period, I visited the laboratory several times and kept in contact with a number of its members. When I first arrived there, I was told that the laboratory had grown quite popular among "anthropologists." I was already the second person interested in studying it.[12] During my six-month stay, I participated in research on population genetics and worked on a project concerned with the typing of Y-chromosomal markers for populations living in the Sinai desert.[13]

Lab P was run by a distinguished geneticist, Professor Svante Pääbo, and it attracted scholars from all over the world to conduct their research there. There were around 20 members, including three technicians. The various projects were clustered in five groups: the RNA Editing Group (concerned with how genetic information may be edited during transcription), the Genome Project Group (concerned with the large-scale sequencing of nuclear DNA and with the technologies to achieve it), the Ancient DNA Group (concerned with DNA studies of extinct species), the Theory Group (concerned with statistical models and computer programs that assist the analysis of the data produced by other groups) and the Population Group (concerned with genetic lineage and comparisons between different human populations). All groups met separately on a weekly basis with the professor, where each member presented his or her work of the previous week and planned further experiments. Since I was especially interested in human genetic diversity, I only attended the Population Group Meetings. I gained insight into other projects through personal conversations, and through the weekly seminars and other in-house meetings where all groups assembled.

Shortly after I had arrived at Lab P, all projects were evaluated. This was an ideal opportunity to acquire an overview of research conducted in the laboratory and to have a first glimpse of the members' projects. These evaluations happened on a yearly basis and were initiated by the professor. In the course of one day, each laboratory member, apart from the technicians, had to present his or her results. Two professors were invited to discuss and evaluate the projects. Jointly with the professor of Lab P, the guest professors would draw up a report with recommendations. During the breaks, we discussed differences and similarities between the projects. It appeared that, apart from three relatively new projects, everybody was working on mtDNA. One could say that, in addition to the professor's good reputation it is mtDNA that binds the laboratory members together.

A Population Group meeting: who is Anderson?

On November 15, we had a Population Group meeting. All members presented their work to the professor. Michael Käser was the last in line. His project was concerned with the !Kung San.[14] He was studying and comparing mtDNA sequences of 25 !Kung San samples in order to answer the question whether the !Kung San could be seen as "a constant population": a population of constant size, which was based on hunting and gathering. During his presentation, he referred to *Anderson*[15] a couple of times. He would state at which position of Anderson the !Kung San would look different from other populations. For one part of the sequence he had studied (about 400 base-pairs), he referred to more than 10 mutations from Anderson.[16] Towards the end of the meeting, I asked him: "What exactly is Anderson?" He explained that Anderson refers to the first publication of a human mtDNA sequence. Because it was the first sequence published, it became the reference. Käser had aligned to Anderson all the sequences he was studying to learn about the positions of mutation. When, for example, he referred to position 311 of the sequences, he was referring to that position in the Anderson sequence. Pääbo, the professor, added to this that Anderson was based on European DNA and that one could, therefore, say that it was biased. Because if one compared the !Kung San with Anderson, one would find more mutations than if a European population were compared. A discussion evolved about this bias in the reference sequence. Pääbo continued that it would not be possible to avoid a bias. You would always be closer to one population or the other. It might be an idea, he added jokingly, to take the Neanderthal as the new reference, since it is equally far away from all humans.

The reference sequence: Anderson

In an interview with Pääbo I asked him some more questions about the reference sequence and about his idea that the Neanderthal would be more neutral than Anderson.[17] In response to my request to tell me a little bit more about Anderson and how it was proposed, he started laughing and explained more fully.

> It wasn't proposed. It was determined. Now, that's a difference between constructionists and geneticists. [laughter]. I mean to say it was the sequence that was sequenced, right. [laughter]. It was the first complete human sequence that was ever determined. It was done in Cambridge. And the first author's name on that paper is Anderson. [laughter]. This was in 1981. It is a composite though, because different parts of that 16.5 kb [kilo base-pair] count for different individuals. It's not even from one person! But the control region is from one person. It had come to be the reference. Because when people determined sequence number two, they had

of course compared that to the first one; since it was already there. And now 4000 sequences later, we still compare to that [first] one. It is of course totally arbitrary. We could take any other sequence as the reference. Anderson is just a convenient convention. Because everybody knows the sequence that you're referring to. But it is of course a British sequence. Or, certainly European, but probably British, because it was done in England.

The difference between "constructionists and geneticists" recounted above underlines that Anderson did not require a "central actor" to become a standard.[18] Anderson became available and turned out to be a convenient convention. Once published as a scientific paper, it started to move between laboratories and to establish its standardized quality in various places in the world. It thus became a convenient convention. When we came to talk about the implications of its Britishness, Pääbo explained that:

> One should of course keep in mind that Anderson is not a golden standard. But if we were to take the Neanderthal sequence, now that would be the perfect reference. It is not a Eurocentric one. And from the analysis we are doing now, it is equally far away from all humans. But it would also be a kind of political correctness, I guess. Kind of purist. Because one avoids the "optical bias" of Anderson. But it wouldn't be practical. In comparison to the Neanderthal, all people would have many shared mutations. These shared mutations are not informative; they would just tell you that people are different from the Neanderthal. They don't tell you how people relate to each other. So one would have to take notice of all those mutations, which are common among humans, before starting to compare humans.

Upon discussing Anderson with other laboratory members, I found that they tended to focus more on what it allowed them to do in practice: "Anderson is just a means to align sequences. From there you compare your sequences and do your analysis." So for them Anderson is a kind of orientation map; a tool to locate the parts of sequences that they are studying. Before addressing the issues of bias and ethnocentricity raised in this interview, let us first have a further look at these practicalities and at what is involved in aligning other sequences to a reference genome.

Anderson: differences and similarities

Based on the idea that humans are basically similar in their genetic material, aligning another sequence to Anderson means to specify the sequence fragment of interest in terms of beginning and end, by positioning the target sequence on the human mtDNA map. Since geneticists are usually interested in small fragments only, this procedure involves tracing the highest matches between the nucleotides (building blocks) of both sequences.[19] So the first practicality of Anderson in laboratories is its functioning as a *locating tool*.

A second practicality is that of Anderson as a means of communication. As Pääbo had put it, "is a convenient convention." Since geneticists are committed to using it and since it is publicly known and accessible, it allows them to specify the results of comparisons, as either similar to or different from Anderson, to other colleagues in the field. Moreover, it accommodates genetics with a nomenclature and a system of how to refer to differences between sequences. It provides a template of the numbering of mutation sites according to the reference sequence. It is in this sense that Anderson could be seen as a piece of communication technology, providing geneticists with a protocol of how to refer to the results of their studies. However, media always come with a message. While facilitating communication, technologies also structure the nature of that communication at the same time.[20]

The talk about the Neanderthal sequence pointed to this quality in Anderson. From the interview, it became clear that as a potential reference sequence the Neanderthal would probably be less biased and maybe less Eurocentric, but it would be less convenient. Humans would look too similar to one another in the comparisons with the Neanderthal. This would "obscure" the differences between humans. Hence, Anderson is not just a locating tool: it also co-determines the comparison, namely a comparison in terms of differences. Geneticists could, in fact, take any sequence as the reference. It does not matter to them whether the reference is European or African as long as it optimizes the detection of differences between the sequences aligned to it. These sequences can then be labeled in terms of mutations *from* the reference. Hence Anderson is a convenient convention not only because it provides population geneticists with a nomenclature but also because it allows them to focus more on the differences and less on the similarities between humans. In communication between scientists, similarities between human sequences move to the background and the differences become the pieces of information transmitted.[21] Therefore, a third practicality of Anderson is its function as a difference-producing technology.

In practice, however, Anderson assists in the production not only of differences but also of sameness. Neither similarities nor differences are self-evident. Why is this? Sequencing DNA is not always straightforward. In practice, the end-result of sequencing may contain some ambiguities. In addition to the familiar letters A, C, G and T, which stand for the DNA building blocks, the sequence may contain other letters, such as N or M.[22] These letters indicate that the visualizing technology could not decide upon which nucleotide it should be: the signals may have been too weak or there might have been more than one signal at the same time. There may be some hints, such as N may be either

G, T or A but not C; or M can be C, A or T but not G. From there onwards, the decision is in the hands of those conducting the experiments. Aligning such a sequence to Anderson, geneticists may decide to leave the ambiguities open, usually by inserting a question mark in the sequence. But they may also edit the target sequence based on experience or arrived ideas in genetics concerning sites that are not likely to vary from the sequence. In a conversation about Anderson, a statistician who was involved with the compilation of a mtDNA databank in Lab P raised this issue with me. She stated: "I always wonder whether people correct their sequences to Anderson, but then I don't know anything about sequencing."[23] In response to my question as to why she actually wondered about that. She said: "I was wondering, because it seemed to me as if the Anderson sequence [i.e. similarities to the Anderson sequence] is found more often than other sequences, and I can imagine no reason why this should be the case. But of course, I first of all have to quantify this guesswork by simply counting the occurrence of the Anderson sequence." This account brings to the fore a fourth practicality of Anderson, namely as a technology of sameness. This quality usually disappears in written sources. Ironically enough, the Anderson paper (the paper in which the sequence was published) reported itself on one of this editing work. The paper stated that: "Nucleotides 10, 934–5, 14 272 and 14 365 were ambiguous and their identity assumed to be the same as in the bovine mtDNA sequence."[24] So, when determining the Anderson sequence, the geneticists also corrected parts of that sequence to an already existing sequence, namely that of a bovine. As a tool for making sense of ambiguous sequences Anderson assists the production not only of differences but also of similarities. It can thus be seen as the "normal" sequence contributing to both similarities and differences.[25] Moreover, the different practices of Anderson addressed here indicate that a reference sequence does not enter the research after the production of target sequences, but during that. Anderson is, therefore, implicated in such sequences and co-determines what they look like.

Anderson in the laboratory

When other sequences are aligned to it, Anderson is practiced as a communication technology within a scientific community and as a normal sequence, contributing to genetic similarities and differences. What about theory? How does this involve theories about genetic inheritance and the understanding of genetic proximity and distance? If so, does this come after alignment?

In the interview when talking about Anderson as a composite made of tissue from different individuals, Pääbo was explicit about one part of it in particular, namely the *control region*. He was quick to state that this part came from one person. Why is it important for Lab P to know this and what kind of information is conveyed by it?

The control region of mtDNA comprises about 900 of the 16 500 base-pairs of the genome and is mainly non-coding DNA.[26] The bulk of this DNA does not code for proteins and does not play a role in the energy-yielding metabolism of the cell. So it is considered "free-to-mutate" (at random) at a much higher rate than the rest of the genome.[27] This feature is important in comparisons between individuals and, consequently, in studies of genetic lineage. The interest of geneticists in differences is theoretically motivated. Mutations are *the* source of information in studies of genetic lineage. The higher the mutation rate the greater the amount of information generated. Shared or non-shared mutations, and the specific ways in which they are distributed within a population, give hints about the presence or absence of genetic lineage.

The presupposition of random mutations in the control region leads to another presupposition, namely that the accumulation of mutations in different populations takes place at the same rate. This randomized and synchronized occurrence of mutations is an important tool in population studies. It is known as the *molecular clock*. This clock is assumed to tick equally fast in all populations.[28] Conversely this means that the number of mutations helps to reconstruct not only lineage but also descent or population history: a higher number of mutations suggests an older population, a lower number a younger population. Populations with a lower genetic diversity are said to have descended from populations with a higher diversity. By estimating the mutation rate (the time between two mutations), geneticists can date historical events, such as the moment of population divergence.

Therefore, random mutation and equal mutation rate contribute to an understanding of differences in terms of proximity and distance between populations. These joint features of the non-coding mtDNA make the control region interesting for geneticists working on genetic lineage and the history of populations and, in consequence, for Lab P. It is, therefore, important for this particular laboratory to know how this specific part of the reference sequence was constructed. As indicated, the very means for establishing lineage are mutations. However, mutations are not an individual phenomenon. They only exist in comparisons. In mtDNA-based genetics, mutations are mutations *from* Anderson. Given the interest of population genetics in the control region, it is this specific part of Anderson that is performed as the "normal" sequence.

The pivotal role of the control region was evident in a photocopy of the Anderson paper that I obtained from one of the laboratory members. In the section of the paper containing the sequence, only the non-coding region had been highlighted by a previous reader. This specific interest has implications for what Anderson is to Lab P. When laboratory members referred to Anderson, or talked about aligning their sequences to it, they were referring in particular to this non-coding part. Similarly, when laboratory members talked about "the Neanderthal sequence," they did not mean, as it became clear to me later, the entire sequence. Matthias Krings did not sequence the whole genome of the Neanderthal. He sequenced just one part of the control region, also called *hypervariable region I*. From the perspective of Lab P, it is obvious to refer to that part of the Neanderthal sequence as the *mtDNA sequence*.[29] The interest of that laboratory was evidently in that part of the sequence; consequently the hypervariable region required no further specification.[30]

From this we can conclude that alignments are theoretically invested. When other sequences are aligned to it, Anderson helps simultaneously to identify other sequences and to produce the very information (i.e. mutations) for comparing them.[31] Given the special interest paid to mutations in studies of populations and genetic lineage, aligning sequences to Anderson does not just mean *reference*, but also *difference* from it. As has been shown, the non-coding mtDNA, in particular, enables comparison based on differences, indicating that Anderson is especially enacted as the control region in the research of Lab P.

Anderson: the British sequence

According to geneticists and to the members of Lab P, Anderson is a convention. It is the first sequence and it works for their purposes. The fact that it is British does not affect how they construct genetic lineage. So one could say that Anderson is not a problem for geneticists. Yet during various talks and during my interview with the head of Lab P, a problem with Anderson's Britishness was articulated. Its "optical bias" and "ethnocentricity" were mentioned as being among its attributes. On another occasion, Pääbo stated: "It's a shame that people forget about the origin of the sequence. They tend to *naturalize* it. They would speak of mutations *from* Anderson, but one could also reason the other way round."[32] This indicated that Pääbo was pleading for situating the sequence in the context in which it was produced. He would like geneticists to be aware of the genetic or racial bias of the sequence: its Britishness and Eurocentrism, which results in a bias. He furthermore argued against its *naturalization* in genetics.

To investigate the issues raised by Pääbo, I suggest we have a closer look at the sequence as a piece of technology. The sequence is a text, which with little effort appears on the computer screen. However, this text was made in a particular time, a particular place and a particular practice. As a text, it can be seen as a "frozen moment" of these constituent parts.[33] To investigate the bias and naturalization of Anderson, let us consider two sides of the coin called naturalization. First, I want to have a look at the kind of nature "brought home" in the sequence.[34] What kinds of practice, technology and tissue, and what kinds of alignment between these, enabled its sequencing and have made it a text? What can we learn from that about the bias in the sequence? Second, I will address the question of how it is that this locally crafted object can travel so smoothly between laboratories to become a *naturalized* tool. The fact that the sequence is a text that can be downloaded from a database does not, as such, explain the naturalization.[35] I will, therefore, examine the practice of theory as a possible site and investigate how that assists the naturalization of the reference. To be sure, naturalization is first examined as a *homing in of nature*, as a way of producing a fit between an object of research and the fabric of the laboratory, and then as the rendering *natural* of a locally produced object by its future users. To do so, I will take you back to the early 1980s, to the era in which Anderson was produced.[36]

Anderson: sequencing

On April 9 1981, *Nature* published a key article on mtDNA. This paper, entitled *Sequence and organization of the human mitochondrial genome*, was accompanied by two other articles giving some further analyses of mitochondrial RNA and it was introduced by a review article entitled *Small is beautiful: portrait of a mitochondrial genome*.[37] The key article opens with the following sentence: "The complete sequence of the 16 569-base-pair human mitochondrial genome is presented."[38] Fourteen scientists presented the complete sequence of human mtDNA for the first time. The map of this genome soon became the reference genome, idiomatically referred to as Anderson. Moreover in the article it was stated: "The DNA sequence was derived mainly from a single human placenta mtDNA preparation, but some regions were determined on HeLa mtDNA."[39]

With hindsight and from the perspective of genetics, 1981 was before the advent of PCR; this revolutionary copying technique was developed in the late 1980s and only became a routine-like technology to allow amplification of DNA from tiny amounts of tissue in the 1990s.[40] In order to understand how people worked with DNA in the early 1980s, I would like to present an account by the population geneticist Mark Stoneking. Although he was not involved in the making of Anderson, his account will serve as an example of how cell material was retrieved, handled and analyzed. Stoneking conducted research on mtDNA in the early 1980s. At that time, he was working together with Allan Wilson and Rebecca Cann. Together they published another key paper in the field, in which the first common ancestor of humans was dated as having lived in Africa

200 000 years ago. This common ancestor has acquired the tantalizing name of mitochondrial Eve (mtEve) or black Eve.[41]

During 1997, Mark Stoneking was on sabbatical leave and decided to spend a year in Lab P. He especially engaged in the Population Group meetings and contributed to the discussions and work conducted by the group. I interviewed him about mtEve and about working with mtDNA in the 1980s in general. Referring to his Ph.D. project in the laboratory of Allan Wilson, he stated:

> Initially I wanted to do Australia, because earlier work indicated that Australian mtDNA is somewhat divergent. But it proved to be impossible to get the samples that we needed from Australia back in the United States. You know in the 1980s, just before PCR, so to do mitochondrial DNA studies we had to purify the mtDNA to its homogeneity.[42] And to do that, we couldn't do that from a blood sample, because you don't get enough mtDNA from blood. So, we have these tissue samples, sort of a placental tissue. So that puts a lot of constraint on what sorts of population you want to get samples from. And it was just impossible to get Australian Aboriginal placentas. But our contact in Australia was a trained student who got to New Guinea to do fieldwork. He ended to be an extremely valuable colleague. Because even though he was in the Highlands of New Guinea, he was able, over the course of two or three years, to arrange a collection of almost a 150 placentas from different parts of New Guinea; to ship them out to us in California and keep them frozen. So they arrived in excellent condition.[43]

Race: homing in nature

To sequence DNA in the 1980s was not an easy task. Geneticists' work had been under a number of constraints. Before the introduction of PCR-based methods, retrieving enough DNA from blood to enable it to be studied was difficult.[44] Although cloning technologies, such as the recombinant DNA technology which enabled the cloning of the DNA, were already established in the 1970s, the tissue was crucial to the success of sequencing.[45] Large amounts of DNA were needed. Unlike blood, placenta is rich in DNA and was considered most *convenient* for (a DNA-based) human genetics.[46] To relate this back to Anderson and the tissue to provide the sequencing samples, given the technical possibilities for cloning DNA, the tissue was also indispensable for its success. Mark Stoneking's account highlights problems in retrieving placentas. In particular, obtaining placentas of people in far-off places was not an easy job. It was a matter of having the appropriate networks.[47] People from New Guinea, for example, may prefer the ritual of burying placentas to other rituals, such as rituals of doing science.[48]

In addition, as his remark indicates, to "bring home" the placenta in a condition that was suitable for use in the laboratory was not as self-evident. It involved putting the tissue on ice within a limited amount of time and keeping it frozen until it reached the laboratory for preparation. It also meant that the placentas had to be shipped quickly before the ice melted.[49] It could, therefore, be said that the crucial role of placentas in DNA studies in the 1980s and the difficulties involved in collecting placentas from all over the world are also involved in Anderson. Whereas placentas from other parts of the world were difficult to collect and ship, collecting British placentas might be simply a matter of going to the next hospital.[50] Children are born and placentas are at scientists' disposal, so to speak. Even though working with British placentas was arduous, it did not require a transnational organization and coordination of activities. From this perspective, it could be said that the sequencing of Anderson was itself based on "a convenient convention," namely the convention of working with placentas because they are rich in DNA, and the convenience of using the tissue at hand (i.e. easy to retrieve).

What does an emphasis on technology and tissue mean for the questions raised above about the bias and Eurocentricity of Anderson? The nature homed in for Anderson in the early 1980s involved a particular handling of DNA. We have seen that cloning DNA in order to study it co-determined which tissue could be considered for sequencing. While the organization of scientific work, such as having access to hospitals and clinics to collect the placentas, or knowing somebody who does fieldwork in other parts of the world puts constraints on and opens up possibilities for whose tissue is to be studied. To take these practices into account means that Eurocentrism and the bias of Anderson is an effect of technology. Race is here practiced in the form of how to handle DNA in a pre-PCR era. In such practices, geographic proximity and distance, the work involved in acquiring good tissue and using the technology and tissue that was at hand became integral parts of scientific routines and co-determined how the sequencing was done and whose tissue it involved. In this particular case, race is an activity of inclusion and exclusion firmly embedded in the routines of genetics and the technologies of cloning and sequencing DNA.

Anderson: the tissue

The 1981 publication of Anderson indicate that the sequence was based on cell material of two human individuals: placenta and HeLa cells.

But what is HeLa? Through a search in Medline I hit upon hundreds and thousands of references to journal articles.[51] Since 1952, HeLa appeared to be broadly used even in papers that were published throughout the 1990s. Also abstracts available from Medline held no more promise of information than the

Anderson paper itself. The papers all stated that HeLa was used, but not where it could be found nor what it was. It was in my *Penguin Dictionary of Biology* that I traced HeLa. Very simple. There it was. "HeLa cell: Cell from human cell line widely used in study of cancer. Original source was Helen Lane, a carcinoma patient, in 1952."[52] So HeLa is an acronym for Helen Lane and it refers to her immortalized cells, the HeLa cell line. Just like the placenta, HeLa cells seem to be a convenient source for DNA, for they come in large numbers. But the specific combination of cell materials used for Anderson, namely placental and HeLa cells, indicates that Anderson is not just a composite. It is not just based on two individuals' cell material, but on material stemming from two women. How, then, should we understand this specific combination of tissue? How, and does the sex of the tissue matter with regard to Anderson?

Un/sexing the tissue: practicalities of homing in

In the interview with Mark Stoneking, I asked him about the problems of retrieving DNA from blood: "Does this mean that before PCR you could only look at tissue from women?" Mark replied: "Right. We did only placental tissue. It turns out that the placenta, I mean, strictly speaking the placenta is a fetal organ. If the fetus is a male, strictly speaking, it's a male tissue. But right, if I'm looking at the mtDNA, it's the mtDNA of the mother."

From this we learn that placentas may be male, but mtDNA is female, namely from the mother. Consequently, the mitochondria of a male fetus and its mother are interchangeable and similar. Hence the sex of the tissue is not an issue for geneticists studying mtDNA; males as well as females inherit and thus have mtDNA. Which makes the combination of tissue in Anderson – placenta and HeLa – seem even more striking, so let us take a look at what kind of information it may contain.

What does it mean to state that: "if I'm looking at mtDNA, it's the mtDNA of the mother," even if the tissue is masculine? This hints at the theoretical understanding of mitochondrial inheritance. As stated above, mtDNA is inherited via the mother only. Males *do* inherit mtDNA but cannot pass it on to their offspring. This particle is "woman made," revealing maternal lineages. Based on this, one could say that the combination of tissue in Anderson seems to suggest a *fit* between mtDNA (revealing maternal lineage) and the tissue (derived from women's bodies).

Mark Stoneking's "sexing" of the placenta seems to point to a different reading. His statement that the placenta is a fetal organ, its sex dependent on that of the fetus, emphasizes the triviality of sex for mtDNA studies. Considering this and the technological possibilities for the sequencing of DNA in the 1980s, the tissue in Anderson could just as well have come from a liver (which is also known to be a DNA-rich tissue), and its source might just as well have been a male. Yet the availability of placentas for scientists to work with cannot be disentangled from a long history of medicalization of the female body, of birth-giving in Western medical practice and of reproductive technologies.[53] Whereas organs of humans tend to be difficult to retrieve in general, a placenta and a cell line have become institutionalized through such practices and may literally function as a *resource* for scientific research.[54] The circulation of such "resources" accounts for their convenience in a laboratory setting. The authors of the Anderson paper, in fact, thank a colleague for "a gift of HeLa cell mtDNA."[55] The placenta used for the sequencing of Anderson was already available in the laboratory. It was cloned and studied for another purpose by Jacques Drouin, a colleague and co-author of the Anderson paper who was based in the same laboratory. The placenta was "described" in a paper written by Drouin, and although the paper does not include details about the origin of this placenta, it is clear that it was retrieved from a hospital or a clinic. In the paper it is stated that: "Human placentae were obtained at term from normal or caesarean section deliveries and put on ice within 30 min." It is also indicated that one of these placentas had been the source for Anderson. "A collection of recombinant clones has thus been obtained using mtDNA isolated from a single placenta and is now being used to obtain a complete nucleotide sequence of human mtDNA."[56] Thus both the placenta – in the form of mtDNA clones – and the HeLa cells – in the form of mtDNA – have become available through an exchange within an organized scientific practice.[57]

Taking this information into account, we see that the specific tissue in Anderson can be regarded as convenient in terms of its availability within an organized scientific practice. However, raising the question about a possible *fit* between a "gendered" theory of mtDNA inheritance and that of the tissue applied demonstrates that convenience is normatively charged. It builds on power relations – in this case gender relations – which exist outside of the laboratory. Even though the sex of the tissue was made irrelevant and was not actively performed,[58] the practices that existed outside the Cambridge laboratory did not cease to matter. Even more, one could say that these practices have become what Donna Haraway has called "frozen moments" built into the reference sequence. They may not be relevant now, but this may change.[59] Thus, in addition to race, sex is also involved in Anderson. We will see below how making sex and race irrelevant is connected to a process of naturalization.

Moreover, the analysis here shows that naturalization is not merely a process involving technology (i.e. how to clone DNA) but also a practice of how scientific work is organized. The availability of tissue is embedded in such practices and co-determines whose tissue is to be used. The nature brought home in Anderson was dependent on materialized gift structures and on the practical organization of scientific work.

We will now shift focus and direct our attention to the second understanding of naturalization, namely the rendering natural of a locally invested object such as Anderson. How Anderson became naturalized, detached from the localities and practices it involved, and enabled to circulate between laboratories, will be issues of concern. To address this we will take a second look at HeLa.

Separating the tissue from its origins: locating Helen Lane

My search for information about HeLa did not stop at the *Penguin Dictionary of Biology*. From a colleague, I received an email message in which he suggested that Helen Lane was black.[60] He had seen a television documentary about a black woman whose cell material had been immortalized in the early 1950s. "Didn't this have implications for my analysis of race and Britishness?" he asked me. This prompted me to start asking various geneticists who HeLa was, but without success. During a telephone conversation with Allan Bankier, the second author of the Anderson paper, he told me that he did not exactly recall either the origin of the placenta or that of HeLa. But he did remember that they had materials of "black" individuals in their laboratory and stated: "At that time these issues were not so much addressed and we were not after an individualized sequence. Our aim was a consensus sequence that everybody could work with."[61]

I received some information about HeLa from the Dutch Cancer Institute in Amsterdam, indicating that Helen Lane may be one of the many synonyms for the woman whose cell material became the cell line.[62] This information included a paper by Howard Jones, a physician who examined Helen Lane, alias Henrietta Lacks. The paper, entitled *Record of the first physician to see Henrietta Lacks at the Johns Hopkins Hospital: history of the beginning of the HeLa cell line*, indicates how difficult it was to grow a cell line successfully in the early 1950s. "The project [of making a cell line] appeared to be a failure until Henrietta Lacks walked onto the stage."[63] She had a specific cancer of the cervix which grew so fast as to facilitate the cell line, so Jones explains. Although the paper gives information about the age of "Helen Lane," the number of children she had and a clinical diagnosis, there is no reference to her color. She died six months after diagnosis and ". . . in terms of Mrs. Lack's birth date, the tumor is some 75 years of age and probably immortal," it is stated.

It was in a paper dealing with population genetics that I found the HeLa cell line addressed in terms of color and origin. The paper, written by Rebecca Cann, Mark Stoneking and Allan Wilson on the subject of mtEve, described the HeLa cell line as "derived from a Black American." In this study, 148 samples from different geographical regions were compared and Anderson was used as one of the compared sequences. Whereas Anderson remains indeterminate in terms of descent, "Helen Lane" and 17 other black Americans were not only qualified as black Americans but were also regarded as "a reliable source of African mtDNA."[64]

The ir/relevance of race: or technologies of naturalization

This story about HeLa suggests that the answers to the questions about race and the Britishness of Anderson are not straightforward.

The HeLa mtDNA samples have travelled from the Johns Hopkins Hospital in the USA to Cambridge, where Anderson was sequenced. Although it is not clear where the placenta came from, a part of the sequence was based on HeLa, which points to Anderson's multiple origin in the DNA. Therefore, the Britishness of Anderson is not so much in the DNA but rather in where and how it was produced. This underlines that Anderson's Britishness is a product of scientific practice. It is a product of organized scientific work and technology that assisted its sequencing.

In the attempts to locate HeLa, the simultaneous presence and absence of "Helen Lane's" color comes to the fore, indicating the relevance and irrelevance of race in genetics. My search took place in 1998 in a European context. This is relevant because, whereas at that time in the USA a large debate was going on about HeLa and its value for scientific research, and about the fate of Henrietta Lacks and her children, who were actually not informed about the origin of the cell line, this information was virtually absent in the European context. The ongoing American debate makes clear that the cell line is thickly entwined with race, class, gender and sexual reproduction.[65] The absence, or rather the inaccessability, of this history in the quest described above is instructive of the relevance and irrelevance of race in genetics. For most geneticists Helen Lane has lost her racial identity through becoming a cell line. Her value for genetic research lies in her rapidly reproducing cancer cells.[66] Also, the producers of Anderson did not mention her racial identity in their paper, testifying to its irrelevance for the reference sequence. As one of the authors, Alan Bankier, has stated, the makers of Anderson were not after an individualized sequence. Instead their aim was to present the scientific community with a "consensus sequence," a standard technology. However in other contexts, such as a population study by Cann and coworkers, Helen Lane's color was seen as highly significant, allowing geneticists to regard her mtDNA as African.

Racial identity is, therefore, pivotal for some scientists and irrelevant to others. However, as part of Anderson Helen Lane lost this identity, not only for the producers of Anderson but also in the localizing work of Cann and colleagues.[67] In the latter, Anderson was taken to be an individual sequence with a European (British?) origin. In this paper, it did not function just as the reference for determining other sequences, it also appeared on a genealogical tree as one of the sequences analyzed. The legend that accompanied the genealogical tree and the table containing the information found, stated: "The numbers refer to mtDNA types, no. 1 being from the aboriginal South African (!Kung) cell line (GM3043), no. 45 being from the HeLa cell line and no. 110 being the published human sequence".[68] Number 45 is placed next to a black square indicating African origin, while number 110 (despite its "hybrid nature") sits next to a square, indicating European descent. Paradoxically, this treatment of Anderson, namely as an individual, gives hints about how it operates as a technology. It gives hints about how it has become a *naturalized* technology in studies of genetic lineage. I will argue that also in this capacity (i.e. as a reference genome) Anderson, just like HeLa or any other sequence, is treated as if derived from one individual. This suggests a procedure through which local practices, technologies and homed in nature are ironed out once Anderson started to move from one place to another. For this purpose we will move back to the 1990s, to Lab P, and take a closer look at the practice of theory. As indicated above, theories about mtDNA inheritance allow geneticists to reconstruct lineage between humans and trace back the origin of their genetic material. The way the mtDNA of "Helen Lane" was treated as African is an example of this in action.

Whose mitochondrial DNA?

Reading publications of Lab P about mtDNA, I noticed that the inheritance of this genome was addressed with a certain degree of caution. It would be described as transmitted "almost exclusively from mother to child."[69] Putting it in these terms suggests that mtDNA could also be inherited from the father. Alerted by this, I started to ask around the laboratory. Lab P is predominantly a mitochondrial laboratory, so I assumed that most of its members had to deal with this phenomenon in one way or another. It turned out that most of them considered mtDNA to be exclusively maternally inherited. Thus paternal contribution was not taken into consideration. Why, then, was it put like that in papers?

One day I raised this question again. A number of us were in the tea room. One members, Hans Zitchler, started to explain that bi-parental inheritance had indeed been demonstrated in mussels[70] and that this had raised questions about whether it could also be the case for mammals. The major question in mammals was whether the mid-piece of the sperm enters the egg cell during fertilization. If so, it would give rise to paternal co-inheritance of mtDNA. Some experiments, he continued, had been conducted on mice. One major problem was that male mtDNA is not easy to detect in embryos, since there was less compared with that contributed by the mother. The mid-piece of the sperm contains only 50 mtDNA molecules, whereas the female cell may contain up to 100 000 copies. The solution they found was to back-cross the mice, a special technology that increases the amount of any DNA contributed paternally. Whereas the first study argued that paternally inherited mtDNA could be detected,[71] a second study came to a different conclusion.[72] Zitchler was especially convinced by the results of the second study. He recalled a talk he had had with the author of the (second) paper, Kaneda. It was during a conference, late at night in a hotel room, where Kaneda explained to him how they had performed their experiments and what the results were. Zitchler reconstructed some of the steps of those experiments for us on the whiteboard and stated that, after having talked with Kaneda, whom he respected a lot, he could no longer be convinced that mtDNA inheritance was bi-parental.

A couple of months later in the course of a get-together, another laboratory member, Valentin Börner, gave a talk about the paper written by Kaneda and colleagues. This get-together, dubbed the Journal Club, is an in-house meeting where members take turns reviewing a paper which may be of interest to colleagues and thus brings it to their attention. At the time, Börner was preparing a review paper concerned with bi-parental inheritance,[73] and, also, he was convinced by the experiments of Kaneda. Not so much because of a lack of paternal mtDNA contribution. It appeared that paternally contributed mtDNA in (back-crossed) mice could be detected after fertilization, but that it was eliminated at a later stage. The fact that paternal mtDNA could be detected at an early stage was found convincing because it meant that the failure to detect this contribution in the later development of an embryo was not a result of insensitivity of the technology used.

However, during the discussion, questions were raised. Pääbo especially found the results "alarming." If paternal mtDNA participates during fertilization, he stated, this might have implications for what "we are doing here in the population group." He made clear that what puzzled him was whether recombination in human mtDNA should be taken into account. That would affect estimations of genetic lineage. As a result of the meeting, Lab P turned its attention to its mtDNA databank to trace possible indications of recombination.

The practice of theory

The previous section has shown that the theory of mtDNA inheritance did not remain stable following the sequencing of Anderson in 1981. A number of scientists became interested in reassessing mtDNA inheritance.[74] Although most members of Lab P were not convinced of paternal contribution, the professor expressed his concern about this, specifically for population genetics. How should it be understood? Why should it be of special concern to Lab P?

Should there be paternal contribution, the major advantage of mtDNA for population genetics, namely that it stems from *one* parent only and therefore does not recombine, would be jeopardized. This one parental inheritance allows for population studies that are based on "simple" statistical models.[75] Thus lack of recombination and rather simple models assist in estimations of genetic lineage. This combined advantage provides an easy system through which similarities and differences between humans can be interpreted as proximity and distance, producing a (mtDNA) track back into history, and leading to the ultimate and mundane parental figure mtEve. The stakes in this ancestor are articulated in the conclusion of one of the papers in which paternal mtDNA contribution was "confirmed" in mice. There it is stated that: "Paternal inheritance of mtDNA also means that mtDNA phylogenies[76] are not exclusively matriarchal. . . . In humans, estimates of the most recent common ancestor of all contemporary maternal lineages could thus be younger than a previous estimate of 200 000 years old, which is based on mtDNA comparisons."[77] Bi-parental inheritance of mtDNA thus has implications for estimations of ancestries and for how lineages are established. Even a small percentage of recombination (e.g. 0.1–1%) would have a profound mixing effect over a period of 100 000 years, which may be 5000–6000 generations. In terms of ancestors, the concept of mtEve itself would become problematical. It would be more appropriate to speak of EveAdam or AdamEve. Moreover, what is at stake is how to interpret similarities and differences between sequences. As stated above, mutations are seen as pieces of information. To interpret this information, however, the time needed for a mutation to occur has to be estimated. It is only under the condition of the molecular clock, at a random and equally high mutation rate in all populations, that the estimates hold and can be used to establish genetic lineage. The occurrence of recombination would introduce a second difference-producing system and would render the link between the occurrence of change in sequences and time more complex. The major problem introduced by recombination is that differences between sequences are not detectable as being the result of either mutation or recombination. Geneticists would then know that both occur and that they have to be treated by different interpretative models. The problem is that they cannot be distinguished as being either of the two. Recombination would then introduce an individualized system of similarities and differences. Occurrence of both difference-producing systems would jeopardize the concept of the molecular clock, blurring the understanding of how humans relate to each other via mtDNA.

Lab P is interested primarily in the control region. This part of the mtDNA sequence is highly variable between individuals and is thus an interesting site for population studies. A case of bi-parental inheritance would have the largest impact on this specific part of the sequence. Whereas a gene in the coding part of the mtDNA sequence would be transmitted as a "package" unchanged, it would either come from the father or the mother. Where parts of the non-coding region came from would be more difficult to trace.[78] Recombination cannot be detected: it simply occurs and has implications for analyses. Since geneticists have failed to find a case of human bi-parental inheritance in humans, models for interpreting genetic differences have not changed and human mtDNA is considered to come from one parent, namely the mother.[79] However, because of this debate, Lab P leaves the possibility open, as expressed in its publications. For as Pääbo put it to me: "But then, you can never know for sure, can you?"[80]

The controversy over mtDNA inheritance is, however, relevant for my analysis of naturalization. I, therefore, suggest we have a brief look at how Anderson and more specifically at how the control region of the reference sequence is compiled.

Anderson: whose mitochondrial DNA?

In the Anderson paper, the "control region" category does not exist. The paper is primarily concerned with locating and interpreting genes, the replication of the DNA strings and the transcription of the mtDNA genome, rather than with non-coding DNA. What is the control region exactly? Where does it begin and end in any mtDNA sequence? When I asked for clarification in Lab P, I was told that to question where it begins and ends would be kind of "academic." "It would be a debate about definitions. In practice, you just make sure that you are looking in the control region."[81] Though this may be the case in laboratory work, it did not help me to align the categories applied in Anderson to that of population genetics, that is to the control region. So I went back to the Anderson paper.

Reading the paper while drawing the different gene positions and calculating the distance between them, I tried to get an overview of what could be called the control region in Anderson. My aim was to find out which tissue was used for sequencing that particular part. Although the paper is ambiguous about which tissue was used for the different parts, it seemed to be more clear about two fragments. They both involve the control region. The text implied that HeLa mtDNA was used to sequence the D-loop, a 680 base-pair fragment. The D-loop contains so-called *promoters*, which initiate the replication of the mtDNA strings.[82] Another fragment, say to the right of the D-loop, was sequenced based on placental mtDNA. This fragment was also of interest to the makers of Anderson because it contains control signals for the transcription of the mtDNA genome.[83] Both these parts of the genome make up a 1100 base-pair fragment situated between two genes, $tRNA^{Pro}$ and $tRNA^{Phe}$. The makers of Anderson stated that, except for the promotors and the control signals, two rather small blocks, this fragment consisted basically of non-coding DNA. So, according to the information in the scientific paper, the fragment as a whole is based on both tissues. The part that was based on HeLa mtDNA, the D-loop, is between positions 16 080 and 191. The second part, based on placental mtDNA, is immediately to the right of the D-loop, between 191 and 580.

Without defining the control region, population geneticists look at sequence fragments somewhere between position 16 000 and position 450.[84] This would then mean that the control region is a composite. However, earlier in the chapter, we learnt that the control region in Anderson was based on one individual. How should we understand this and what kind of information does it contain about the naturalization of Anderson?

Anderson made natural

Anderson is a composite and so is the control region, so the scientific paper suggests. The fact that the control region as a category was absent in the Anderson paper hints at the existence of different practices, the practice of population geneticists and that of the makers of Anderson. This difference can be expressed as a difference between non-coding and coding DNA. The makers of Anderson sought to understand the functioning of the genome and the contribution of its genes, whereas population geneticists seek to understand differences in the non-coding DNA in terms of population history and genetic lineage. While population geneticists treat the non-coding DNA as one category, namely the control region, for the makers of Anderson this fragment consisted of more categories, the D-loop and a second locus where control signals for transcription were identified. It had become clear that these categories were defined by different positions on the genome and seem to have been sequenced on the basis of different tissue (see Appendix II). So how should we understand the practice of population geneticists, where the control region of Anderson is treated as coming from one person? Or, better, what can we learn from that about the naturalization of Anderson?

The debate about mtDNA inheritance addressed above made clear the stakes in maternal inheritance. The current theory supports the single origin of mtDNA. It is inherited from the mother and ultimately from the "all human mother," mtEve. This theory of inheritance makes all individuals comparable in their DNA based on established models of the mutation rate in the control region. The productivity of a theory of mtDNA as inherited from one parent is the very convenience by which the control region in Anderson is taken to be one and produced as one: that is, based on the mtDNA of one individual. This suggests that population geneticists may take their objects too seriously and their practices too much for granted. They take their practices for granted in the sense that they tend to forget about them and about how they are involved in objects. They take their objects too seriously in the sense that they forget about their fabricated nature, and their dependence on the practices that helped to produce them. Objects thus seem to be autonomous and coherent entities that can be found everywhere.

To bring this debate back to why it is that the control region of Anderson is taken to be derived from one person, one could say that the very presupposition about the origin of mtDNA – as stemming from one individual – is doing the work of naturalization for Anderson as well. Both Anderson and mtEve are products of theory and practice. Unlike Anderson, however, the sequence of mtEve does not exist. It is a concept that is dependent on a theory of inheritance and on standardized practices of comparing sequences. It is dependent on standardized methods of analyzing differences and on the universalization of results, in such way as to establish lineages that lead back to that single origin.[85] This very standardized and universalized approach to genetic diversity and lineage has the naturalization of Anderson as its effect. Above, I have shown that in one publication Anderson appeared on the genealogical tree. Not as a composite sequence (a technology) from mtDNA of more individuals, but as a singular unified person. By contrast, in medical practices where genealogy and the concept of Eve are not that relevant, scientists have reported extensively on mistakes in Anderson. They have in a sense *denaturalized* the reference sequence, because it was not in line with their medical findings. In 1999, this denaturalization led to the resequencing of Anderson (see Appendices I and II).[86]

In population genetics, the universal approach to difference that enables mtEve also enables the naturalization of Anderson. In the absence of Eve from which all humans descended, and her sequence from which all humans diverged through the accumulation of mutations, Anderson is performed both as a tool to establish mutations and as an origin from which all other human sequences diverted. The very concept of mutation in this practice establishes Anderson as the origin. To be sure, this origin quality is not to be understood in a genealogical sense. Rather, it is a practical effect of working with the reference sequence on a routine basis, a procedure which is constitutive of Anderson's "naturalness" in laboratories. It is enacted as the materialization of a sequence that has gone missing. The way the practices of the makers of Anderson were taken for granted and the control region was taken to stem from one individual is based on an investment in the concept of a single origin. Therefore, with the help of a universal theory of mtDNA inheritance, Anderson had become the original sequence from which all other sequences descended, "the mitochondrial Eve of modern genetics".

Conclusions

The Diversity Project is dependent on various technologies that have been produced in different places and time. Laboratories are populated by many more-or-less standardized technologies. Technologies tend to obscure their normative content or even the appearance that they have been produced somewhere. They seem autonomous and natural. In this chapter, I have taken one such technology as my example to analyze its normative content and investigate how it has become a naturalized tool in laboratory practice. My object of analysis was a mtDNA reference sequence, the Anderson sequence. This sequence is not just any technology, but a rather central one in studies of genetic lineage and diversity. I have examined the practices, technologies and tissues that helped to produce it, in order to highlight the biases it carried along with it. My analyses have shown that racial and sexual biases are involved in Anderson, and that these are effects of the technologies applied, the handling of DNA and of the organization of scientific work. Biases such as the ones traced here make the questions about naturalization even more urgent.

In the case of Anderson, I have shown that its naturalization is dependent on another object, namely the universal theory of mtDNA inheritance. While universalizing differences to establish lineage, this theory facilitates a treatment of Anderson as being both a technology for establishing differences and the source from which all sequences have derived. My analyses suggest that the problem of naturalization is not really caused by a quality built into a technology itself. Technology can never be neutral and is always produced somewhere. The problem should rather be sought in enhanced alignments achieved in specific practices, which render a technology self-evident. For the problem with naturalization is not only that it tends to obscure the normative content of technology, but also that it helps to essentialize the differences produced by it. For example, nobody carries mutations by her/himself; it is not an essential feature of an individual. It is only in relation to such a standard that individuals can be compared in terms of differences. Even more, in a laboratory setting, target sequences and Anderson become intertwined, and for ambigious sequences sameness has to be established actively. However, naturalization tends to move the technologies and routines involved into the background, and to put differences (or similarities) into the foreground as characteristics contained in individuals and populations. Therefore, in order to prevent the essentialization of difference, we need to take account of technologies, as means to denaturalize the objects they help to produce.

Appendix I
Revising the standard

In June 1999, I started working in Lab F. My identity had shifted and my task was no longer a study of laboratory practice. My contract stated junior researcher population genetics, and I was working on a project concerned with mtDNA diversity among the Dutch. My work consisted primarily of the sequencing of a fragment of the control region for 168 Dutch samples. Peter de Knijff, head of the laboratory and, therefore, my boss, had read my analyses of Anderson and knew of my interest in the control region. One day in October I told him that I would be interested in sequencing HeLa mtDNA. Half an hour later, over lunch, he let me know that a cell line had been arranged from a scientist working in the same building.

Once the cell line had arrived, I extracted the DNA and sequenced a 420 base-pair fragment. Aligned to Anderson, I found two mutations. Since the cell line reproduces so rapidly, we were expecting mutations. The mtDNA we had, while still from the HeLa cell but now 20 years later, was no longer the same as the one used for Anderson. For Peter de Knijff, finding only two mutations was an indication that the site was rather conservative.

Just about in the same week that we had that information an interesting article was published in *Nature Genetics* on a reanalysis and revision of the Cambridge reference sequence for human mtDNA.[87] Four scientists had sequenced the placental DNA originally used for Anderson as well as HeLa DNA. They had found mistakes in vital genes and suggested corrections for the revised reference sequence. In their article they referred to a debate in the early 1990s among geneticists studying genetic diseases related to mtDNA. In one of these articles it was stated "this sequence is a composite sequence of a human placental and HeLa cell mtDNA and therefore is not an appropriate reference standard for such [medical] studies."[88]

I contacted one of the authors of the recent publication, Professor Doug Turnbull, and asked him for further information. In one of his email replies, he wrote the following.

> When we sequenced the HeLa we just used available HeLa cells and thus not the same DNA that Sanger *et al.* used. Neil Howell recently presented this data at a meeting at which Guiseppe Attardi was present. Attardi apparently sent the HeLa DNA to Sanger for the original paper and we are hoping to get some of this DNA and then sequence. A further comparison will then be possible. Do you want to wait until we have this sequence before I send you the Sequence?[89]

So the original placental DNA had been sequenced anew and the same is going to happen for the original HeLa DNA. The now corrected sequence of Andrews and others has found its way to the databank and it has replaced the Anderson for those working in the field of medical and population genetics. This indicates that a reference sequence may go unnoticed in some practices but not in others, and that naturalization is very much a product of the kind of practices to which technology is aligned.

There is another point to this story. Bruno Latour has argued that questioning of stable scientific facts requires the building of "counter laboratories."[90] It seems that in the era of the HGP and of postgenomics, technologies are enabling in this respect. The sequencing of Anderson was an achievement in the 1980s and is made light work nowadays. Hence, the development of technology may change the stability and status of standards more rapidly than it had before. Moreover, this example also shows that the questioning of standards does not necessarily have to be the work of – let us say – social scientists. Specific practices and specific localities within the life sciences may also call for revisions of standards.

Appendix II
Revisiting HeLa

While working on the manuscript for this publication, I wanted to expand on HeLa: its "African American" origin in relation to the reference sequence and in relation to the issue of race. In that process, I noticed a scientific paper published in 2002 reporting on the first relation. In the paper *A high frequency of mtDNA polymorphisms in HeLa cell sublines*, Herrnstadt and colleagues describe the sequencing and analysis of the mtDNA genomes of six HeLa sublines and aligned these to the revised Anderson sequence of 1999 (rCRS).[91] Among the collection of HeLa sublines they have acquired the cell line of Attardi (mentioned in Appendix I). This is the so-called CalTech HeLa cell line, which was used for the sequencing of Anderson in 1981. Similarly to the questions raised in this chapter, namely how is Anderson composed and which part of it was based on which tissue, the researchers were interested in "resolving uncertainties" surrounding the composite nature of Anderson. They provide new facts that are highly interesting in themselves and exemplary of how race is treated in genetics. I shall, therefore, take a further look at their work.

In the scientific paper, the researchers observed that the compilation of the Anderson sequence had become rather inaccessible.

> The sites for which the bovine sequence was used were reported in the original publication, but not those for HeLa mtDNA. Unfortunately, there does not appear to be any existing laboratory record that provides this missing information (A. Coulson, personal communication).[92]

One piece of information however was available in the Cambridge Laboratory:

> The original placental mtDNA was cloned as a series of 23 Mbo I restriction fragments for subsequent DNA sequencing. However, three of these fragments (numbered 2, 5, and 8) yielded few or no transformants, and they are the most likely mtDNA regions for which HeLa mtDNA could have been used.[93]

One of these fragments, fragment 5, encompasses parts of the D-loop (nucleotide 15 591–16 569). The researchers, however, excluded this site as being contributed to the Anderson sequence by HeLa mtDNA. The CalTech HeLa mtDNA differed at five different positions from the Anderson sequence.

Therefore, in contrast to what I put forward above, the control region appears to be based on placental mtDNA only. So where does that leave my analyses? I will first elaborate on that point and then will address what is happening to the issue of race in this study and that described in the related paper of Andrews and his colleagues.[94]

As I have described above, while studying the Anderson sequence and how it was compiled, I have first tried to gather my information from the laboratories that I was studying. I then turned to one of the scientists involved in the making of the Anderson and finally ended up relying on the original publication for acquiring any information on how it was composed and which parts were based on which tissue. The various elaborations on the D-loop on the basis of HeLa mtDNA in the original paper seemed to indicate that HeLa was used for that very part in the sequence. This, combined with a further elaboration on one part of the control region while referring to placental mtDNA, made me deduce that the control region was a composite, based on HeLa and placental mtDNA. The fact that scientists were asking similar questions about the composition of the sequence 20 years after it had been introduced, and were trying to provide evidence for the loci of HeLa and placental mtDNA in the reference sequence, is in line with my conclusion that this information had gone missing, and it underlines my analyses that the composition of the sequence, including the control region, was rather unclear and could not be presupposed from the available information. It is for this reason that I have left my initial analyses unchanged and that I contend that the discussions about the naturalization of Anderson on the one hand, and of the control region in relation to controversies over mtDNA inheritance, on the other, are still relevant and valid. They are exemplary of scientific practice and the technologies taken for granted.

With the resequencing of the placental tissue in 1999 and the HeLa sublines in 2002, Anderson had ceased to be a consensus sequence and had become an individual reference sequence instead. It is no longer a composite but is based on placental mtDNA only. The resequencing of the placenta allowed scientists to eliminate 10 errors that had occurred through ambiguities in the sequence and the use of bovine mtDNA, but also other so-called *errors* from the use of HeLa cells. One "error" in particular is interesting in terms of race. Referring to the fragment that was based on HeLa mtDNA, Andrews and his colleagues stated:

> The only *error* that we were able to explain using the HeLa sequence was that at nt [nucleotide] 14 766 (T versus C, respectively). The revised CRS [Anderson] mtDNA belongs to European haplogroup H on the basis of the cytosine at 14 766 . . . [emphasis added].[95]

Thus, replacing the fragment that stemmed from HeLa by placental mtDNA had resulted in one nucleotide change, from T to C at position 14 766. However, by correcting this so-called error, the sequence had not only become an individual reference sequence, rather than a composite, but it had apparently also turned from an African sequence into a European one. The sequence that was produced in 1981 was *assumed* to be European through the quotidian *deleting* work of science, which made the normative content of the sequence inaccessible. The new Anderson is *made* European through the *editing* work of science, which turned the HeLa mtDNA into an error. This correction is striking if we take the placental mtDNA into account.

There might, of course, be a number of reasons for the decisions taken for the revision of the Anderson sequence. The two articulated reasons in the papers cited here are the making of an individual sequence and the eliminations of errors. Both are interesting in terms of naturalization. The move from a composite fabricated kind of sequence to an individual sequence contributes to the naturalization of the revised Anderson. That is, it contributes to the idea of a mtDNA sequence existing out there in the world from which other sequences diverge: mutate. As we have seen, to achieve that, the sequence had to be made into a European one, containing a C instead of a T at 14 766. However, this very belonging to a particular group is problematic in the case of Anderson. In the 1999 paper the following is stated about these peculiarities: "There are an additional seven nucleotide positions at which the original CRS is correct and which represent rare (or even private) polymorphic alleles."[96] This then means that the Anderson sequence contains polymorphisms that are uncommon among Europeans or any other population and some others which are even said to be private (i.e. not yet found among human individuals). These polymorphic sites are maintained in the revised sequence. It is stated that "The rare polymorphic alleles should be retained (that is, the revised CRS (RCRS)) should be *a true reference sequence* and not a consensus sequence." [emphasis added].[97]

Whereas these divergences from other Europeans are preserved and contribute to its individuality, the "HeLa error" is corrected as to make it fall into a European population. Thus this combined individuality and Europeanness contribute to the further naturalization of the revised Anderson sequence, in terms of a real existing individual sequence. The sequence will therefore cease to raise questions that might refer researchers back to its erratic history full of practicalities and decisions. There is, however, one other peculiarity that could still generate such questions.

Whereas scientists had decided to correct the "HeLa error" as well as a number of sequencing errors that they had found, they preserved one error in the revised Anderson. At one particular locus, the makers of Anderson in 1981 had reported two Cs (nucleotide positions 3106–3107), while the resequencing of placental mtDNA in 1999 showed that it had actually to be one C (at nucleotide position 3106). Yet the revised sequence will continue to encompass that error. "The last suggestion represents a compromise between accuracy (correcting the numbering to account for the single C residue at nt 3106 and 3107) and consistency with the previous literature. We believe that renumbering all of the previously identified sequence changes beyond nt 3106 would create an unacceptable level of confusion."[98] This very choice, motivated by practicalities, may function as a "noise" reminding the users of the reference sequence of its fabricated nature. Since the second C will not be found in any other human sequence, that lack may become a sign for the practices that helped to produce it. It may thus denaturalize the revised Anderson. However, this choice also contradicts the scientists' wish to present "a true reference sequence". The sequence contains an extra C, which might as well have stemmed from a different tissue, since it is not in the placental mtDNA. This, in turn, makes the choice to correct the "Hela error" even more striking since there is no practical reason whatsoever to "correct" that according to placental mtDNA. Given the extra C, the sequence cannot be said to be an individual sequence. Yet it had become a European one owing to the very correction of the "HeLa error". Consequently, the revised Anderson had not become a "true reference sequence" because it is in agreement with the mtDNA of one individual and, particularly, because it falls within a population group, namely European.

Notes

1. Anon. (2001a,b).
2. Krings *et al.* (1997).
3. See note 16 in Chapter 3.
4. The results were eventually published in the journal *Cell* (Krings *et al.*, 1997).
5. It should be noted that the "sequencing" of so-called ancient DNA (DNA of extinct species) is very much indebted to other parts of the experiments, such as DNA extraction and purification. Since geneticists do not know what they are looking for nor what it should look like, and since there is no other sequences with which it can be compared, they have to conduct many experiments in order to purify the target DNA from other genetic material that may have become part of the cell tissue.
6. The location of chromosomal DNA in the nucleus was, in fact, already known in late nineteenth century (Kevles, 1985).
7. In mtDNA, TGA instead of UGG codes for tryptophan (Trp), ATA instead of AUG codes for methionine (Met) and AGA and AGG signify termination codons instead of arginine.
8. The numbers are rough and difficult to estimate for nuclear DNA, but that does not have implications for the point I am trying to make here.
9. Watson *et al.* (1991: 446–447).
10. For this feature of the mtDNA genome, see von Haeseler *et al.* (1996: 135). In Chapter 5, I compare mtDNA to the Y-chromosome because the Y-chromosome

also has this quality: absence of recombination, which makes it useful for historical/evolutionary genetic studies.

11. Kevles and Hood (1992a).

12. It seems that not only laboratories have come of age, as Karin Knorr-Cetina (1995) had put it, but also laboratory studies themselves. After the initial problems that scholars of science and technology encountered in conducting laboratory anthropology, it is proving ever easier to find support among scientists for this type of study. The story goes that it is quite fashionable among scientists to have an ethnographer studying their laboratories. Similar to commercial advertising and the artistic efforts put into advertising by biotechnology companies, it seems that ethnographers contribute to the prestige of (non-commercial) laboratories. On the procedures of acquiring access to laboratories in the late 1970s and early 1980s, see Latour and Woolgar (1986); Law (1994); on the key role of commercial advertisements in science and analyses of the work advertisements do within and outside the sciences, see Haraway (1997a).

13. This project will be elaborated in Chapter 5.

14. The exclamation mark in the name of the !Kung San refers to the specific language of this "population," namely the click language.

15. Whereas scientists would speak of "Anderson" in colloquial talk and in the laboratory, in their publications they usually refer to the sequence as CRS (the Cambridge Reference Sequence). In this chapter I will side with the every day practice of genetics and continue to refer to the sequence as the Anderson sequence.

16. Positions 93, 129, 153, 167, 172, 209, 212, 214, 223, 230, 234, 239, 243, 260, 266, 278, 291, 294, 311. I thank Sonja Meyer for forwarding this information to me.

17. The interviews conducted were put on tape. For the sake of clarity and readability, the transcripts presented here do not contain utterances such as "ehh" or information about pauses. Also half sentences or sentences stretched out are rephrased in the transcript to make them more comprehensible. For the purposes of the analyses I conduct here, this extra information is not considered of importance. The interview with Professor Svante Pääbo was conducted on 4th February 1997 in the Laboratory for Evolution and Human Genetics in Munich.

18. See Timmermans and Berg (1997: 275).

19. On the use of the human (nuclear DNA) genome, which is being sequenced in the HGP, see Fujimura and Fortun (1996: 164–165). They analyzed the difference between genetic similarity and "homology" and argued that, in contrast to similarity, homology is theoretically invested by theory about human evolution.

20. For this notion of technology, see Donna Haraway (1991a). She argues that "Technologies and scientific discourses can be partially understood as formalizations, i.e., as frozen moments, of the fluid social interactions constituting them, but they should also be viewed as instruments for enforcing meanings." On the ideological content of computer technologies and of formal representation, see Star (1995a).

21. For a good example, see Sajantila *et al.* (1995); see also Torroni and Wallace (1995).

22. In a report on mtDNA analysis in forensics, Kevin Sullivan and others state that some of the sequences "are less clear cut," and that some of the sites "were scored as 'N's' by the automated software." In another paper, Romelle Hopgood, Kevin Sullivan and Peter Gill (1992: 89) reported how they got rid of ambiguities in their sequences: "Resultant sequences were initially compared using DNASTAR

'COMPARE' computer program that enabled most sequence ambiguities to be resolved."

23. Sonja Meyer's concern with this may be understood in the context of her work, namely compiling an mtDNA databank. The published and unpublished sequences she collected for this purpose consist of complete sequences as well as sequences containing many ambiguities. For this databank see, Handt *et al.* (1998); Burckhardt *et al.* (1999). On current discussions of errors in this mtDNA databank, see Arnason (2003); Dennis (2003).

24. Anderson *et al.* (1981: 458). Also, Gyllensten *et al.* (1991: 255) reported on a study of mice mtDNA, stating: "The C57BL sequence has been corrected relative to Bibb *et al.* [the mtDNA reference for mice, AM] at positions 15 823 and 16 119."

25. Charis Cussins (1998: 67) described normalization "as a means through which new data . . . is [sic] incorporated into pre-existing procedures and already pre-existing objects."

26. The control region consists of non-coding DNA, also called hypervariable regions I and II, as well as two small "conserved blocks" of the sequence that *control* and initiate the replication of the molecule and the transcription of the genetic information it contains. In Anderson *et al.* (1981: 458), a larger part is identified as containing little "open reading frames" (genes) namely a fragment of 1100 base-pairs.

27. It is not clear how much higher the mutation rate of the non-coding DNA is in relation to coding mtDNA. Early studies suggested a ten times higher mutation rate (Wilson *et al.*, 1985: 387). More recent studies suggest a variable mutation rate between fifteen and twenty times higher (Pääbo, 1996: 494).

28. In a handbook of genetics (Watson *et al.*, 1991: 444) it is stated: "The concept of a molecular clock is based on certain assumptions. One assumption is that sequence differences between the same genes and proteins in two species have accumulated since the species diverged from a common ancestor. Furthermore, the rates at which the differences accumulate are assumed to be equal. If this were not the case the clock would run at different speeds in two species. For a more cautious approach to the concept of the molecular clock, see Wilson *et al.* (1985: 388).

29. See Krings *et al.* (1997: 23).

30. One could say that in the context of Lab P, Anderson is both a synecdoche and a metonym. Anderson, the name of a geneticist, comes to stand for the sequence, its accomplishment and application as the reference (a synecdoche). The control region, however, is a part of that sequence that stands for the whole, namely mtDNA in general (a metonym).

31. On this double quality of scientific objects, see Lynch (1990); on the politics of standards, see Star (1991, 1995b).

32. This was at a conference we both attended at the Max Planck Institute for the History of Science in Berlin in July 1998: *Postgenomics? Historical, Techno-Epistemic and Cultural Aspects of Genome Projects.* The optical bias refers to how *genetic trees* are assembled. In the genealogical tree of Rebecca Cann and her colleagues, this optical bias is most visible. There Anderson takes a place among almost exclusively European "individuals," whereas Africans, for example, are placed on the other end of that tree (Cann *et al.*, 1987: 34).

33. Haraway (1991a: 164).

34. See Karin Knorr-Cetina (1995: 146) for an elaboration on this feature of laboratory science. She argues that "laboratories allow for some kind of 'homing in' of natural processes; the processes are 'brought home' and made subject only to local conditions of the social order." See also Latour (1987).

35. See Chapter 2 for an example of this argument in the context of DNA fingerprinting.
36. In par with Annemarie Mol, I want to apply the notion of *locating* as a strategy to situate the problem of naturalization *somewhere* in scientific conduct; see Mol (1985, 1990, 1991a).
37. Borst and Grivell (1981).
38. Anderson *et al.* (1981: 457).
39. Anderson *et al.* (1981: 458).
40. See Rabinow (1996a); Chapter 3 describes PCR and how it is handled in laboratory practices.
41. Cann *et al.* (1987). Mark Stoneking from this grouping came to work in Wilson's laboratory in 1981.
42. The determination of the presence, constancy (stability) and frequency of DNA.
43. Interview held on 11th March 1997 in The Laboratory for Evolution and Human Genetics in Munich.
44. Blood was at that time collected on a global basis for determining blood types and protein variations. "By the mid-1960s, a large number of clear cut biochemical variations were known, including more than a dozen inborn errors of metabolism arising from probable enzyme deficiencies, and so were numerous haemoglobin and blood-serum protein variants" (Kevles, 1992: 15).
45. Anderson was sequenced according to such a technology. It was cloned with the help of the bacteria *Escherichia coli* (*E. coli*) (Drouin, 1980: 15). Moreover, Drouin conducted the work for this paper in the laboratory of F. Sanger (at Cambridge), a co-author of the Anderson paper and whose name is especially connected to the development of sequencing techniques and of recombinant DNA in the early 1970s (Sanger and Coulson, 1975; Sanger *et al.*, 1977).
46. The so-called blood-type genetics (i.e. looking at blood groups in studies of genetic diversity, admixture and migration history) was introduced during the First World War by the Polish couple Hirszfeld. Blood was widely researched throughout the twentieth century not only for reconstructing genetic lineages but also in the context of family diseases (Kevles, 1985: 202–204). A nice representation of the enormous impact that PCR has had on the number of mtDNA sequences (i.e. of parts of the control region) that were determined is offered in a paper from Lab P: Handt *et al.* (1998). In this paper, a diagram entitled "Accumulation of HVRI and HVRII (hypervariable region I and II) sequences during the last 15 years," showed that approximately 100 sequences had been determined before 1990, when PCR was published as a method; from that year on, more than 4000 sequences became available for HVRI. In this paper, a map of the world is also given, indicating from which parts of the world the current collection of sequences was derived (i.e. which populations are represented (Handt *et al.*, 1998: 126, 127)).
47. See also Clarke (1995); Anderson (2000).
48. About rituals of burying placentas in New Guinea, see Strathern (1991: 128). On rituals in science, see Jordan and Lynch (1992). Cann *et al.* (1987: 32) gives an example of how geneticists circumvented the problem of collecting placentas in "far out places," using placentas of black Americans as a source for mtDNA representative of Africans.
49. Clarke (1995: 195–198).
50. For an example, see Cann *et al.* (1987: 32). For their study, Cann and her colleagues received 98 placentas from US hospitals.
51. Medline is a database for scientific publications. This database is compiled by the National Library of Medicine, Bethesda, Maryland, USA.

52. Thain and Hickman (1996).
53. Mol (1985); Martin (1987); Oudshoorn (1994); Ginsburg and Rapp (1995); Franklin and Ragoné (1998); Pasveer and Akrich (1998).
54. The case is different in animal genetic research. Organs of animals, such as rat liver and beef heart, were studied by Anderson and his colleagues (1981: 458). In a study of bovines, for example, liver and brain tissue were used as a source for mtDNA (Olivo *et al.* (1983: 401); large amounts of mtDNA was obtained with PCR from liver, heart and kidney in mice (Gyllensten *et al.*, 1991: 256).
55. Anderson *et al.* (1981: 464). Organs of animals, such as rat liver and beef heart, were studied by Anderson and his colleagues (1981: 458).
56. Drouin (1980: 16 and 15). I thank the second author of the Anderson paper, Dr Allan Bankier (at the MRC Laboratory of Molecular Biology, Cambridge), for pointing out to me that they had received the mtDNA clones from Dr Jacques Drouin, and for providing other information about the sequencing of Anderson.
57. See also Warwick Anderson's (2000) study of Kuru brains as objects of exchange among scientists, and on how the Kuru brains oscillated between a gift and a commodity status.
58. Similarly, Marilyn Strathern (1988) argued that gender is not inherent in the objects as such, even when women are the objects of exchange. Rather, it is the activity of and the ends of transactions that may become gendered. See also Hirschauer and Mol (1995), who argued that gender may become mute and requires to be performed actively. Chapter 5 will investigate further the stability and instability of sex differences.
59. On the shock and debates that followed, once it became clear to scientists in 1966 that the origin of the HeLa cell line was a black American woman, see Landecker (2000).
60. I thank Ruud Hendriks for his attentive reading of a previous version of this chapter and for bringing the point to my attention.
61. Dr Allan Bankier, telephone conversation on 17th November 1998.
62. I thank Professor Professor Piet Borst and Suzanne Bakker of the Dutch Cancer Institute (Antoni van Leeuwenhoek Ziekenhuis) for providing me with this information.
63. Jones (1997: 227).
64. Cann *et al.* (1987: 32).
65. The way Anderson is treated as an unmarked category, a universal standard, is strikingly similar to the way the HeLa cell line had been such an unmarked and universal category. In both cases, the issue of race had been erased in more or less effective ways. On HeLa see Landecker (2000).
66. On how this quality had become a problem, see Landecker (2000). The HeLa cells are apparently growing sufficiently fast as to contaminate other cell cultures and, in fact, complete laboratory spaces.
67. Cann *et al.* (1987).
68. Cann *et al.* (1987: 34).
69. In Pääbo (1996: 493). It is expressed in a similar way, for example, by Pääbo (1995), Sajantila *et al.* (1995) and Salem *et al.* (1996).
70. In this study it was shown that both maternal and paternal mtDNA is transmitted to male offspring, but only maternal mtDNA to female offspring. This led to the concluding working hypothesis, that ". . . sex is controlled by the mother's genotype, which also controls, pleiotropically, the fate of the sperm's mtDNA. Eggs determined to be females also become able to prevent the entrance of the sperm mitochondria, destroy them after entrance, or suppress the replicative

advantage of the male mtDNA. Alternatively this advantage may be present only
in eggs determined to become males" (Zouros, 1994b: 7466). On bi-paternal
inheritance in the blue mussel, see also Laurence and Hoekstra (1994); Skibinski
et al. (1994); Zouros (1994a: 818).
71. Gyllensten *et al.* (1991). The first study by Gyllensten in the 1980s on paternal
inheritance in mice did not show evidence of paternal DNA. This was explained in
terms of technology. The technology was not considered sensitive enough to detect
a possible contribution (Gyllensten *et al.*, 1985).
72. Kaneda (1995).
73. Börner *et al.* (1997).
74. For a literature review of this debate, see Ankel-Simons and Cummins (1996).
This article is, as the title indicates, on the side of bi-parental inheritance, and it
provides a large corpus of literature from different fields, such as anatomy,
genetics, embryology and popular science.
75. Handt *et al.* (1998: 126).
76. The objective of phylogenetic studies is to reconstruct genealogical ties between
organisms and to estimate the time of divergence between organisms since the last
shared a common ancestor.
77. Gyllensten *et al.* (1991: 257). Soon after this publication, this conclusion found its
way into a handbook of genetics (Watson *et al.*, 1991: 446–447), where the results
were presented as an achievement of technology (i.e. PCR cloning possibilities).
78. Wilson *et al.* (1985: 384). There it is stated that ". . . the consensus at present is
that recombination does not happen in animal mtDNA except perhaps in the
displacement loop [major part of the control region]." This is not so strange given
the fact that the coding region of the mtDNA genome contains only very small
fragments of non-coding DNA, see Anderson *et al.* (1981: 457).
79. There are, however, single cases that show the occurrence of mtDNA stemming
from both parents, for example Bromham *et al.* (2003).
80. In 1999, the debate over bi-parental inheritance was reopened, Smith and Smith
(1998, 2002); Hagelberg *et al.* (1999) (the claims in this last article were
withdrawn in August 2000).
81. Dr Matthias Krings, personal communication in December 1998. The word
"academic" is generally applied by scientists in a pejorative sense, as a contrast to
analysis, based on data, and empirical work, based on experiments.
82. Replication is the DNA copying process necessary during cell division. In the
genome, the D-loop is situated between position 16 080 and 191 (Anderson *et al.*,
1981).
83. Transcription is the process of producing RNA (ribonucleic acid) from coding
DNA, as part of producing proteins.
84. For the purpose of the mtDNA databank, compiled in Lab P, it was decided to take
fragments in two regions: positions 16 001–16 408 and 1–408. This decision was
based on available sequences (Handt *et al.*, 1998).
85. Essential to such standardized analyses of genetic differences is the concept of the
molecular clock. A clock ticking equally fast in all individuals and which helps
geneticists to estimate the time for mutations to occur. See Chapter 5.
86. See Andrews *et al.* (1999). On discussions about mistakes in Anderson among
physicians, see, for example, Howell *et al.* (1992).
87. Andrews *et al.* (1999).
88. Quoted in Marzuki *et al.* (1992: 1338). See also Howell *et al.* (1992).
89. Doug Turnbull, personal communication 11th October 1999.
90. Latour (1987).

91. Herrnstadt *et al.* (2002); Andrews *et al.* (1999).
92. Herrnstadt *et al.* (2002: 20).
93. Herrnstadt *et al.* (2002: 26).
94. Andrews *et al.* (1999).
95. Andrews *et al.* (1999: 147).
96. Andrews *et al.* (1999: 147).
97. Andrews *et al.* (1999: 147).
98. Andrews *et al.* (1999: 147).

5

The traffic in males and other stories on the enactment of the sexes in studies of genetic lineage

Introducing the argument

What is genetic sex, and how is it enacted in studies of genetic lineage? These are the main questions put in this chapter. Genetic sex is hardly an issue in population studies interested in human histories. However, in the laboratories, one may find samples indicating male or female, and published papers contain accounts of women's migration history and that of men. This suggests that sex does matter. But where can it be located? It will be argued that, rather than a stated message in the DNA, in laboratory practices the sexes are performed as various things. However, this diversity tends to be subsumed and differences between the sexes tend to be essentialized. I will examine how that is done in the context of mitochondrial DNA (mtDNA) and Y-chromosomal research and show that this requires a specific treatment of DNA, namely as both a technology and a resourse for studying the history of populations.

Human geneticists know the sexes as XX and XY. Critics of this binary scheme, especially feminist scholars, have argued that to differentiate on the basis of XX and XY is to fail to pay any attention to culture. My aim in this chapter is to show that neither of these approaches takes into account the practices of genetics. Difference between the sexes is neither a natural quality embodied in individuals nor a cultural additive. Rather, it is an effect of interfering practices where the sexes are deemed relevant. Thus instead of taking culture as the fact after the biology I view culture as part and parcel of biology and examine sex differences in the practices of genetics. I will view the relevance and irrelevance of the sexes in a laboratory context where experiments are conducted, and in published papers where the data is analyzed and put in the context of population history and genealogy.

The relevance and irrelevance of the sexes

How 153 male samples lost their sex

On January 18 1997 I took the night train from Amsterdam to Munich to continue my participant observation. Among other things, my luggage contained 153 blood samples. All the samples had been taken from males during a large-scale survey on heart disease in the 1980s. All males were 35 years of age and were living in the small Dutch town of Doetinchem. I picked the samples up at Lab F.[1]

I was taking the samples along at the request of Maris Laan, a member of Lab P. Laan was working on a project in the field of population genetics concerned with population history and the spread of agriculture in Europe. She studied this by looking at "linkage disequilibrium"[2] on the X-chromosome. To avoid complexities induced by recombination, she decided to look at male DNA only, since males carry only one X-chromosome and not two as females do.

When she visited me in Amsterdam in July that year she brought along half of the DNA extracted from the blood samples. We placed the two boxes in my refrigerator and the accompanying forms on my desk. The forms referred to the samples as Du208, Du209, etc. and indicated the DNA concentration of each sample. This was determined through mtDNA PCR products and visualized on agarose gels of which a Polaroid picture was included. The samples were to be delivered to Lab F as a return favor for making the blood available.

However, something strange had happened to the Dutch male blood samples that had travelled to Munich earlier that year. Not only because they had come back as DNA samples, but also because they had lost one of their qualities along the way. The very quality that made them move from Leiden to Munich in the first place: their sex.[3] In the forms, they were referred to as "Dutch blood samples" and the DNA concentration was not determined on the basis of the X-chromosome but on that of mtDNA amplification. The samples were still qualified as Dutch but no longer as males. So why did these samples lose their sex or how could it be enacted?

Sex and sexual differences are not stable. As Stephan Hirshauer and Annemarie Mol have argued, they have to be performed actively.[4] This indicates that they may become irrelevant altogether. To consider the relevance and irrelevance of the sexes, let us first have a look at feminist studies of science and then at the account introduced above.

Feminist studies of science

Feminist scholars have put a great deal of effort into showing that science is, just like any other practice, sexualized. They set out to show that it was sexualized in terms of who does the research, revealing a male bias and bringing to the surface the contributions of women in science. Others examined the language of science, providing insight into hierarchies in the designation of agency, and about biases between objects categorized as masculine and others as feminine. Again others considered scientific methods and have argued that these could be seen as masculine. Methods were shown to establish a distinction and a hierarchy between a (masculine) subject of research, namely the scientist, and a (feminine) object of research, namely nature.[5] These approaches led to one basic feminist claim concerning sexual differences. "Sex" can be found everywhere.[6] "You just have to put on gender glasses to see it," as one scholar once put it to me. Once I had been in the laboratories, however, the sexual distinction I found seemed banal, the kind of distinctions that I could have learned about in any other environment. In addition, there was nothing specific to genetic sex in laboratory practices. Yet population geneticists' accounts of human history talk about men and women and their different migration histories. So how does this *materialize* in laboratories and where can it be traced?

The strategy I propose and will follow here is not that of putting on "gender glasses." For the focus is usually set somewhere else and could obfuscate what is to be looked at. Moreover, putting such glasses on metaphorically exposes the wearer to the danger not only of predefining – if not essentializing – the sexes and what counts as sex differences but also of developing a blind spot for the irrelevance of the sexes. Instead, I follow a strategy proposed by Annemarie Mol, a strategy of locating objects in practices. From this perspective, a universal claim, such as that sexual differences are relevant in any kind of practice and, consequently, also in genetics, gives rise to the question "where is genetic sex and how is it performed?"[7] Another strategy suggested by Evelyn Fox Keller is that of counting: "counting past two." Keller has used this approach to draw attention to the diversity in science as well as that in "gender." This numerical and tantalizing practice has an advantage that I would like to emphasize here. A commitment to counting, especially when the sum is more than two, prevents us from taking the binary scheme of biology for granted and, additionally, the practice of counting also involves a risk, namely that of not finding even one.[8] This other side of the coin of counting, "added to" the strategy of locating as suggested by Annemarie Mol, enables both the making specific of a universal claim and of the practices where such a claim does not hold: in this case, practices where sex is irrelevant.

Let us now consider some of the information embodied in the story about the Dutch male samples.

As we saw in the Dutch example, samples do not travel without additional information. This does not mean, however, that all information is deemed important or is even stored in the files. The Dutch samples from Doetinchem, taken from 35-year-old males in the context of a large-scale study of heart disease, lost many of their features in the written forms that accompanied them back to the Netherlands. The very reason why these samples travelled to Munich was because, according to the genetic matrix XX–XY, males carry only one X-chromosome. Once the Doetinchem samples had established their Dutch-ness and maleness through the accompanying information, they became simply Dutch, and they were referred to as such in the accompanying list on their trip back to Leiden. It is remarkable that the sex of the Dutch samples that entered Lab P was deemed pivotal but that it no longer mattered (i.e. it lost any material reference) once these samples had left the laboratory.[9] It indicates that sex may be a temporality performed in locales, and that it is not an essential feature of samples.[10]

Taking the temporality of genetic sex into account, the aim of this chapter is to examine if and how the sexes are enacted in studies of genetic lineage. To do this, we will consider three sites in studies of lineage. These sites will be referred to as the practices of genealogy, of DNA in the laboratory and of genetic lineage. In a way, these are the sites of theory, raw data and analyzed data. It is, however, important to emphasize that differences between these sites are analytical rather than ontological. The differences do not correspond to the classic division between "hand labor" and "mental labor." They rather point to a gradient of technologies, spaces and problem-solving procedures that are more or less important in these three practices. Since studies of genetic lineage are dependent on lineages themselves, namely lineages between laboratories and scientific groups, let us first take a look at how that is done.

The economy of exchange: the traffic in males and other gifts in genetics

The title of this section makes play on the titles of two classic works in the field of gender studies.[11]

In November 1996, at a weekly Population Group meeting of Lab P, Maris Laan reported that she did not have enough European samples necessary for her project.[12] Laan is in charge of all samples that enter or leave the laboratory and she had noted that although the laboratory had many samples of populations from all over the world, there were hardly any Europeans among them. All she had found were samples of Swedes, Finns, Estonians and Samis. Confronted with this problem, especially since her project involved studying European population history, she raised the point during the meeting. We started to brainstorm about where the laboratory could ask for samples. A large amount of German DNA samples would soon be available as a result of collaboration with a Dutch med-ical research group.[13] Another possibility, I suggested, would be to ask Lab F in Leiden for Dutch male samples. Having spent some months in this laboratory myself and knowing that they worked with a Dutch control population for forensic purposes, I reasoned that they probably had plenty of DNA.

After the meeting, I called the head of Lab F, Peter de Knijff, and it turned out that they did indeed have a large collection of Dutch male samples and were willing to share part of it with Lab P. The arrangement was that they would give Lab P blood samples and get back half of the DNA extracted from them.

When I visited Lab F in January to pick up the samples, de Knijff gave me some further information about them. Each sample consisted of two blood samples placed in plastic tubes, which together was 10 milliliters of liquid blood per sample. This blood was "taken up" in a buffer (EDTA) to preserve its quality. They were placed in a box and put on dry ice for the forthcoming journey. De Knijff asked me to assure Pääbo and Laan that the sampled population was a good representation of the Dutch population at large, because Doetinchem showed neither a founder effect (i.e. that a limited number of individuals would account for the genetic diversity of its current inhabitants) nor recent admixture, which would be reflected in the genes and distort their representativity of the Dutch.

From February onwards, Laan Maris and I started to extract DNA from these samples. She would refer to our laborious work as "our large-scale extraction." Because of the large amount of blood that we had from each sample, extracting DNA from a series of 30 samples would take us two days, in which there was hardly any time left for any other work. A considerable part of this time was spent in planning the work, making sure that all the chemical solutions and equipment we needed was at hand, including enough pipettes and tubes, plastic bags for disposals, good pencils to mark the tubes according to individuals, different colors according to which step of the extraction had been performed and deciding which extraction protocol would be the most efficient. We first worked with a phenol–chloroform DNA extraction protocol, but Andreas Kindmark, a colleague, suggested using a sucrose gradient–high salt extraction method. He had used this in a medical laboratory in Sweden and told us that it was not only user-friendly compared with the phenol–chloroform method but it also required fewer steps before retrieving DNA. If we were interested, he would contact his laboratory in Sweden and ask for the exact method. Once the protocol has arrived, Laan tried it out for three samples and found that it also yielded an amount of DNA similar to that yielded by our original method, so we changed protocol. Part of the organizational work was booking machines such as the centrifuge, deciding who does what and making sure that we protected ourselves well because we were working with so much blood. During the first steps of the extraction, using gloves, masks and appropriate chemicals to clean our working environment was not so much to avoid contamination of the samples but of ourselves, namely to protect ourselves and others in the laboratory (who make use of the same environment) against contagious viruses, especially hepatitis B. The "laminar flow cabinet" where we conducted the extractions would be packed with rather large pieces of equipment for the first stages of the extraction: bigger pipettes to pipette the clotted blood into the buffer placed in 20 milliliter tubes. Toward the end of the procedure, once the cell material had been separated from the DNA-containing supernatant, the equipment became smaller, the pipetting more precise, the treatment of the tubes more careful, to avoid mixing the centrifuged DNA at the bottom with the solution at the top of the tube. Thus the logistics of doing DNA extraction, which makes up the bulk of the work, are reflected in the treatment of the samples from blood to DNA.[14]

Making lineages in genetics

From the above account, it is clear that undertaking population genetics is dependent on an economy of exchange. The exchange of samples is as valuable as the exchange of (unpublished) data, extraction protocols or research methods.[15] I will first consider the samples.

There are many reasons why collections of samples can be found in a particular laboratory. Samples may be there because they are considered valuable: rare or difficult to retrieve. They may be there because the laboratory happens to know the people who originally had them or, in many cases, they are there simply because they were offered. As Laan's account showed, these gifts show a bias towards "exotic" populations or populations from regions of the world other than Europe. Other samples, as in the case of the European samples, may be in the laboratory because geneticists who came to conduct their research had brought them along. In a sense, the samples just happen to be there for anybody interested in doing a project in population genetics. Samples may also enter a laboratory because of an express demand, originating in an ongoing project, as in the example of the Dutch male samples. The design of Laan's project prescribed that the samples should be European and male. Consequently, the samples received were selected according to sex, and their Euro-Dutch origin was assured since they were from the small town of Doetinchem and not, for example, from Amsterdam, where the population is, given its complex demographic history, much more problematic to categorize as Dutch or even European.[16]

The traffic in samples is very much dependent on a "gift economy." It is dependent on lineages between laboratories or scientific groups. At the same time, once samples, start to move they establish and help to strengthen lineages. In the case of the Dutch samples, a recently initiated collaboration facilitated the gift and the exchange of blood for DNA. Lab P had just started to work with Y-chromosomal markers using the protocols of Lab F. According to conventions, Lab F would be acknowledged in papers for the gift of samples as well as for the marker information, either in terms of a co-authorship or under Acknowledgements. Also, in line with this reciprocal gifting, the head of Lab F, Peter de Knijff, was invited to Lab P to give a seminar on Y-chromosomal research in April 1997 and Lab F was able to add 153 DNA samples, instead of blood samples, to its collection. From the description of the amount of work involved in such extraction, it is easy to see why DNA samples are preferred to blood samples. The procedure of extracting DNA also benefited from gifting protocols: the unexpected gift of an alternative extraction method from one of the laboratory members expedited the laborious work and proved to be friendlier for those carrying it out. The appearance of this protocol from a Swedish medical laboratory hints at another type of exchange between laboratories.

Lineages are not only related to the traffic in samples, technologies or scientific data, but also to the traffic in people. In addition to the exchange of samples and markers, Lab F and Lab P have also exchanged their "in-house anthropologist," a person who knows both laboratories quite well and who contributed to a more informal traffic in Dutch samples, to information about the samples and to communication between the laboratories.[17] Hence exchanging "in-house anthropologists" makes lineage as well. This position is not exceptional but applies also to other visiting researchers in Lab P. Hence the protocol from the Swedish laboratory. Also other laboratory members establish lineage and are part of kinship relations between laboratories, scientific groups and countries. These lineages may be temporary, lasting mainly for the period the researcher is in the lab, or of a more durable kind. Therefore, genetics does not only study lineages, it is also a product of lineages, established through an exchange of people, samples, technologies and methods.

So this is how geneticists *do* lineage between themselves. But, how do they *do* genetic lineage? This is the topic dealt with in the next section. We will first focus on the relevance of the sexes in practices of genealogy. It will become clear that, in these practices, DNA is not merely a resource containing information, but also a technology, involving different systems for doing genealogy.

Archeology of the human genome: facts and artefacts of genealogy

Archaeology of the human genome is part of the title of a paper by the population geneticists Arndt von Haeseler, Antti Sajantila and Svante Pääbo. This paper – let us call it the archeology paper – provided a literature review and argued for the potentials of genetic data in reconstructing human history, especially when the two-sexed model of mtDNA and Y-chromosomal DNA are considered.[18] The paper opened as follows.

> Many of us, especially in our youth, are interested in the lives of our parents and immediate family; then again, as members of a particular group or population, we like to know about the life of our ancestors; finally, as members of the human race, we are fascinated with the question of human origins. . . . However, early humans

left traces of their activities not only in the form of their bones and artefacts. They also passed on to us their genomes. Every genome is made up of about three billion base-pairs, several of which experience mutations in each generation, and, as the way in which these mutations accumulate in populations are influenced by how populations expand, contract, split and merge, the study of genetic variation has the potential to yield a great deal of information regarding our history.

Under the heading "A bit of theory" it goes on.

All individuals have parents, and some individuals have the same parent(s). The consequence of these trivial facts is that as genealogical lineages in a population trace back over generations, they will occasionally coalesce to common ancestors. There will be fewer and fewer ancestors as one goes back. Eventually, all female lineages will trace back through a series of consecutive mothers to one single mother and all male lineages will similarly trace back to a single father, that is, the most recent common ancestors (MRCAs) on the maternal and paternal side. . . . If the demographic history of a population is unknown, it can be reconstructed from the patterns of nucleotide substitution in the genome. DNA sequences from mitochondrial (mt) genome and those from the majority of the Y-chromosome are particularly useful as they are passed on without recombination from mother to daughter and from father to son. Consequently these sequences can be traced back directly to the genealogical maternal or paternal MRCAs. Autosomal DNA sequences, which are inherited through both males and females and occur in two copies per individual, trace back to "biparental" MRCAs that are on the average four time as old as maternal and paternal MRCAs.

Under "The age of the human gene pool," the paper indicated that these maternal and paternal most recent common ancestors (MRCAs) are expected to be found around 200 000 years ago. However, it is not clear to geneticists and to the authors of this paper whether our species (i.e. modern humans) originated at a time close to these MRCAs.[19]

Genealogy and technologies of the sexes

The choice for the term *archeology* in the title of the paper quoted indicates that specific distributions of similarities and differences come with a story about the past.[20] A story about populations. Similar to the treatment of archeological artefacts as records of human history, mutations in the DNA and the way these are distributed among populations are treated as records of population histories. Under the assumption that all populations have one origin, the differences in particular can be read as events in the past, contributing to stories about when and how populations diverged or merged, reduced in size or grew. As the quote indicates, mtDNA and major parts of the Y-chromosome are considered very useful for studying these events, particularly because neither recombines. They are inherited unchanged from mother or father, and so these DNA systems represent the maternal and the paternal line of inheritance, which can be traced back to one ancestral mother and one ancestral father. Before addressing these two systems, let us take a closer look at a how genealogy is practiced in studies of genetic lineage and at the relevance of sex in these studies.

The trivial fact of genealogy mentioned in the archeology paper, namely the fact that all individuals have parents,[21] demarcates an involved relation between genealogy and genetic lineage. From a genealogical perspective, going back in time means to unfold a greater complexity in biological kinship.[22] It makes more and more individuals appear as part of "the family": as ancestors of a specific individual. From the perspective of an individual, this amelioration of ancestors can be represented by the form of a **V**. While the intersection between the two arms of this letter indicates a contemporary moment in time where there is one individual, their divergence points deeper and deeper into history where progressively more ancestors can be located in the space between the two arms. However, the archeology paper contended that there "will be fewer and fewer ancestors as one goes back." This suggests that, from the perspective of genetic lineage, the genealogical **V** should rather be turned upside down, to become a **Λ** instead. At issue here is not an ever-growing family but an ever-shrinking family the further one goes back in time. Yet, how should we understand this type of genealogy? How should we understand the occurrence of a **Λ**?[23]

Although the opening sentence of the archeology paper evoked the idea that population genetics is interested in individuals and in where they come from, its main focus was rather on groups of individuals: a population. Population genetics studies how such groups relate and reconstructs the development of these relations through history. Therefore the object of study, many individuals and not one, explains the **Λ**. The space at the bottom (the largest divergence between the two arms) stands for a group of individuals and a contemporary moment in time. So, *what* is the relevance of **Λ** and *why* does not each of these individuals have his or her own **V**-shaped genealogy? The answer lies in *how* geneticists study individuals and for what purpose.

Geneticists do not study all genetic material but focus on very limited fragments of DNA. Their aim is a study of lineage for particular parts of the DNA. Tracing back the descendent of a particular piece of genetic information take into account only those 'parents' who contributed by passing it on. Others are left out. Doing so, the further geneticists go back in time, the fewer ancestors and the more lineage they can presuppose. Ultimately, so the quote indicates, this genetic information coalesces in two ancestors, a mother and a father. V is hardly ever a topic in population genetics. If V stands for the genealogy of an individual, in the context of DNA it would require study of a large amount of genetic information; by comparison, Λ stands for a specific type of genealogy, one that has one DNA fragment as its topic and so helps to establish genetic lineage. Whereas V is about how the individual is connected *to* predecessors, the Λ is about how individuals are connected to-each-other *via* predecessors. This also indicates that the fact that all individuals have parents gains importance in studies of lineage, which attributes a specific meaning to that very fact. From a genetic perspective, individuals are first and foremost products of sexual reproduction. All parents pass on their genetic material via sexual reproduction to individuals. However, whereas in the V type genealogy, parents themselves are seen as individuals with parents and grandparents, thus permitting the V-shape, this is not the case in the Λ type. Parents or ancestors whose genetic material is not represented in present generations (i.e. the specific DNA fragment under consideration) are left out of the picture. This means that although sexual reproduction and parenthood are pivotal for studies of genetic lineages, focusing on a limited amount of genetic information and being interested in comparing it between individuals, parents become necessary passage points of genetic information. They are not so much relevant as individual males and females but as nodes through which many individuals can be related to one another. This then means that in studies of genetic lineage the sexes are not relevant as male and female parents but as a source of reproduction and therewith as passage points through which genetic lineage is established.[24]

Given this central role of sexual reproduction, it is interesting that the archeology paper stated that mtDNA and the greater part of the Y-chromosomal DNA are especially appreciated for studying genetic lineage, because they escape recombination. From a genetic perspective, recombination and sexual reproduction are interchangeable. The case of mtDNA and Y-chromosome shows that geneticists study the effects of sexual reproduction, namely genetic lineage, through particles that are excluded from sexual reproduction. Therefore, mtDNA and the Y-chromosome are deemed valuable because they are conveyed by a reproductive system through which lineage can be established. Let us briefly ponder these two systems.

Unlike the Y-chromosome, mtDNA is to be found in the cytoplasm and not in the nucleus. Situated in the mitochondria, mtDNA is passed on via the mother only. Males and females inherit their mitochondria and so too their mtDNA via their mothers: that is, via the egg cell. Males have mtDNA but do not pass it on to their offspring; only females can do that.[25] This system of inheritance accounts for a maternal lineage. The Y-chromosome, however, shows a different pattern. Fathers do not pass on their Y-chromosomes to female offspring but solely to male offspring. Only males carry Y-chromosomes and pass them on to their sons. A pattern of inheritance that accounts for a paternal lineage. From the perspective of the individual, however, there are other differences between the two systems. Viewed from the mtDNA approach, there is no difference between males and females. They both have mtDNA. From the perspective of the Y-chromosome, male and female individuals differ. Only males carry this chromosome and they pass it on solely to male progeny. What does this difference between the two DNA systems tell us about the relevance of sex in studies of genetic lineage? The story of mtDNA, in particular, indicates that geneticists are not interested in the sex of the individual. Even though males do have mtDNA, it is considered to be part of the female line of inheritance. Interestingly enough, the archeology paper even excluded males from that system. It stated that mtDNA is passed on "from mother to daughter." This suggests that fathers are analogously not acknowledged for the fact that they carry a Y-chromosome, but for the fact that they pass it on to their sons. Therefore, in practices of genetic lineage, sex is performed not so much as a quality of an individual but rather as a pattern of inheritance. Hence sex differences are not located in the individual but in genetic kinship.[26]

This specific relevance of sex difference can be viewed further if we take MRCAs into account. The paper mentions three categories of MRCAs, one single mtDNA mother, one single Y-chromosomal father, and a third type consisting of many single autosomal "bi-parents," autosomal referring to the 44 chromosomes located in the nucleus and inherited from both parents.[27] Both maternal (mtDNA) MRCA and paternal (Y-chromosome) MRCA are estimated to have occurred about 200 000 years ago;[28] the "bi-parental" MRCAs, however, may be four times older, suggested the archeology paper. This implies that from a genetic perspective our MRCAs do not necessarily have to coincide with individuals or with actual parents. From this perspective, MRCAs can best be seen as partial products of genetic lineage. In line with this, DNA is handled as a variety of technologies which, together with a Λ-type genealogy, contribute to producing those lineages. Moreover, as I have argued in the case of genetic lineage, genetic sex is not performed as a quality of individuals but as a pattern of inheritance or – better – a technology of lineage. Similarly, DNA is not so much treated as an essential feature of individuals but as a technology "embodying" different systems for producing lineage leading to different MRCAs.[29] Taking the mtDNA and Y-chromosomal systems into account, this treatment of DNA can, therefore, be seen as a technology for producing sexualized genetic lineages.

The ir/relevance of sex in laboratory practice

In the course of my participant observation in Lab P, I was working on a project that was trying to compare two "bottlenecks" (a reduction in genetic diversity resulting, for example, from a reduction in population size), one in the Sinai Desert and one in Finland, by studying the Y-chromosome. For this purpose, the Finnish population was compared with that of Sweden and the Sinai populations with those living along the Nile. Studies with mtDNA have shown a reduced diversity in Finns compared with Swedes, and the same has also been found for three Y-chromosomal markers tested.[30] The population in the Sinai Desert looked slightly different. Their mtDNA showed as high a diversity as in the rest of Egypt, whereas the three Y-chromosomal markers showed a reduced diversity. Lab P was interested in testing more loci on the Y-chromosome to explain this difference and to see whether that difference still held when more markers were applied. Abdel-Halim Salem, who was working on both the Finnish population and those living in the Sinai, familiarized me with the project and we continued working on it together.

Before we started working with the markers, Salem drew a map of Egypt to show me where the populations along the Nile and from the Sinai were living. Discussing the faith of the populations of the Sinai, he explained to me that most of the samples we had in the laboratory, except for the samples he (a medical doctor) had collected himself, were assembled in the 1960s by an Israeli population geneticist, Professor Batsheva Bonne-Tamir.[31] These were serum samples[32] and since they were so old, their quality was not always that good. The set of Y-chromosomal markers we were about to use for the Finland–Sinai project were sent to Lab P by Lab F. Salem showed me the set of primers, the "ladder" for each marker, some control samples (tested in Leiden) and the protocol that was already so familiar.[33] I, nevertheless, brought in my own copy of the protocol containing notes and remarks I had made earlier when I was in Leiden. We went through the protocol and talked about how to establish the PCR condition: writing the programs, making the reagents, measuring the concentration of the primers produced in Lab P based on the Leiden primer and testing the markers for a small number of samples from a population called Sawarka (Sinai). Once the markers appeared to work, we extended our work of typing them to more individuals from that population. The strategy Salem proposed was to do one population at a time for all markers and then move on to the next.

After I had finished typing one marker (*DYS 390*) for all the Sawarka samples, I found only two alleles: two fragment lengths.[34] Instead of going on to the next marker, I decided first to compare these results with another population, Jabalya, which was an exception in the Sinai. Previous studies had shown that, contrary to other populations in this region, it showed no reduction in diversity on the Y-chromosome.[35] I was of course curious as to whether that would hold for this marker as well. I was unable to discuss the change in method with Salem since he was abroad, so I took the samples from the refrigerator and started running the PCRs.[36] From the 36 samples that I tested, none of them showed a band on the agarose gel. I then thought that, it might be that the bands were not very strong and that they could nevertheless be detected by the ALF$^{®}$, which is a more precise visualizing technology. So I booked one machine for the next day. However that too was not positive.[37] When Salem came back, I told him about the "Jabalya problem." Although he was at first a little annoyed that I had changed the plans, when I showed him the collection of samples that I was typing he started to laugh and stated: "Now we can tell Svante that we know for sure that the samples are females." It appeared that I had been trying to type the Y-chromosomes of females. We walked over to another part of the laboratory where he showed me a file in which I could find information about the laboratory's samples. It contained different kinds of information, in many cases information about their sex, about when and where the samples had been collected and who had supplied them to the laboratory. He told me that if this file did not contain information about the sex of the samples, I could have a look at his personal file on the Sun computer, where he had stored his raw data, including data about the three Y-chromosomal markers that he had typed earlier. He explained further that with some of the samples it was unclear whether they were male or female, and then he stated "I don't even know if all non-males are females." This is especially a problem of serum samples, because if they fail to work for nuclear DNA, you cannot determine whether this is a result of deterioration of the DNA or because they are females and do not have a Y-chromosome.

Following this episode, we started reorganizing the samples according to sex. We first took a second collection of the Jabalya samples and separated the two sexes in the boxes and then did the same for the other populations. Then I made a list of all the Sinai samples that were known to be males and wrote this information down in my laboratory journal.

Technologies of DNA/technologies of sex

In practices of genealogy, sex mattered as a pattern of inheritance. It mattered in the way it helped to establish genetic lineage. However, in a DNA practice, as a procedure of producing data at the bench, the sex of the individual as illustrated above became a significant part of studying DNA.[38]

Compared with mtDNA, studies of the Y-chromosome are rather new in the field of population genetics. The former has been used extensively since the 1970s.[39] The first population studies on the Y-chromosome, however, appeared in the early 1990s,[40] and it was only in 1995 that a number of Y-chromosomal markers were discussed as being informative for the purpose of population studies.[41] In Lab P, the first Y-chromosomal markers were introduced in late 1995 and the laboratory's first paper reporting work carried out using these markers appeared in 1996.[42] This information reflects the organization of daily work in Lab P and the relevance of sex in doing DNA.

Working on the Y-chromosome, it is relevant that some individuals have a Y-chromosome, namely males, and others do not. Are these then females? Just as in the Jabalya case, the absence of Y-chromosomal alleles was read by

Salem as extra information about the female sex of those samples. This allelic information contributes to a "practice of chromosomes," that is a practice of XX–XY. In this practice, the sexes are performed as presence or absence of the Y-chromosome. However, working with rather old samples showed that this distinction is not "natural." Absence of a Y-chromosomal allele does not necessarily mean that the individual from which the sample was taken was a female. In this case, the sex of the sample is an effect of good or deteriorating DNA. Presence of a Y-chromosomal allele makes a sample into a male sample. Its absence does not make a sex difference. Sex difference could, therefore, be seen as a local and contextual laboratory product, invested by the quality of individual samples and their relevance for a particular experiment. Therefore, in a practice of deteriorating DNA, sex is performed not as a quality of a *sampled individual* but as that of an *individual sample*.

Laboratories reflect the activities carried out in such space, and the organization of the space is often centered around such activities.[43] For example, in the laboratory there are cupboards above each bench containing most of the chemical solutions needed for the specific work conducted at that specific bench, and there are always sets of pipettes, pipette tips and latex gloves within easy reach. The samples are also subject to this type of organization. There is a spatial division between individuals according to population, which are preferably stored in separate boxes. In addition, old and new samples (i.e. serum and blood samples) tend to be stored in separate boxes. Sex, however, did not bring about such a spatial division. Males and females were mingled and placed in the same boxes. So how should we understand a mutual relevance and irrelevance of sex differences? How should we understand the pivotal role of sex for working with Y-chromosomal DNA, and the virtual absence of sex in the organization of work?

Whereas I had problems seeing any system in the numbers assigned to the samples (some series would have unsystematic numbers, such as "101," "7125" or "77&78;" others would have a number and a letter referring to the name of the population such as "B9," "B31" or "B91;" still others would have a number and two letters such as "FB25" or "MB29," indicating males or females of that same population), Salem seemed to have the relevant information at hand.[44] Simply by looking at the containers of DNA he would indicate to me which samples were gathered when; which of the samples were male or female, and, so he told me, he even knew personally some of the people represented by the samples he had collected himself. This information was neither absent nor irrelevant, even though it was not visible to a newcomer. This also applied to information about the sexes. Having worked much longer with the samples, Salem could be said to have embodied that information. My knowledge of the samples was limited, so

I had to mobilize other practices of knowing the sexes by consulting the written records, the Sample file and the raw data in the computer. For Salem, these practices were already part of the letters and numbers that were written on the cups. Moreover, since he had been engaged in collecting some of the samples, other *repertoires* of enacting the sexes were at his disposal. These repertoires consisted not only of written records and previous experience in the laboratory but also of an anatomical way of knowing the sexes in the field. In such a way, for example, the presence or absence of breasts makes sex and enters the form as such. Moreover, his remark about knowing some of the individuals from whom we had samples indicated yet another repertoire and another practice of performing the sexes.[45] This is a practice in which the sexes are performed as social differences between men and women and where individuals can be referred to as Mr A or Mrs B, the brother of so-and-so or the mother of this or that person. This also made it easier to personalize and sexualize a DNA sample. Salem's knowledge regarding the samples in the laboratory was thus based on an *interference* between different repertoires and different practices where the sexes were performed.[46] I had to introduce another way of establishing the sexes, namely that of creating a visual distance between male and female samples. By making a spatial division in the boxes and drawing up a list of all the male samples in my laboratory journal I created means of transforming these different ways of knowing the sexes that became pivotal parts of doing DNA.

As indicated above, work conducted on the Y-chromosome is rather new in population studies as well as in Lab P. The populations we were typing for the Y-chromosome were first studied using mtDNA. Unlike the Y-chromosome, both males and females carry mtDNA. From the perspective of mtDNA, the sex of the individual is not relevant. Any human sample will do, even those for whom the sex can no longer be determined.[47] Therefore, the storage of samples according to population, or even according to the DNA quality of the samples, can be seen as reflecting a former practice, a practice of doing mtDNA. Although the laboratory was moving away from mtDNA[48] and although Salem as well as other laboratory members had already conducted Y-chromosomal research for which sex did matter, the samples occupied "the same place" as before. The changed practice was not reflected in *how* the samples were organized spatially. Rather, it was operative as a management of different repertoires of performing the sexes. Managing these repertoires revealed an organization of different practices. Whereas anatomical and social practices of performing the sexes, to which Salem had access, were dominant over a practice of written records, where sexual difference did not always appear, the practice of records was dominant over my lack of knowledge about samples. Still further, a practice of DNA deterioration, as in the case of the old samples, was again dominant over the very practice of records and eventually over that of anatomy and social order, if these repertoires were no longer to hand.[49] Hence, while the sexes were absent in the organization of the samples in the lab, the sex of the samples could be performed as an effect of interfering practices and as the management of different repertoires.

The analysis of laboratory work has shown the relevance and irrelevance of the sexes. Whereas sex was irrelevant in a mtDNA practice, in a Y-chromosomal DNA practice it was enacted as a quality of the individual samples. Before addressing practices of genetic lineage, let us take a second look at the organization of samples in the laboratory. This will make clear how much the mtDNA approach was involved in Lab P's collections of samples and how it has affected the gifting of samples.

The relevance of the sexes: sexing the gift

While I was in Lab P, many new samples were coming in: from Russia, Estonia, the Middle East, Nigeria. One set of samples was destined specifically for the Y-chromosome typing project. At one point, I walked into the laboratory and found a box of DNA samples on my bench. Along with it there was a note from the head of the laboratory saying that the Bosnian samples had arrived and asking whether I could see to it that they were stored properly. I was expecting the samples since we had talked about them previously during a Population Group meeting. There it was mentioned that we could obtain Bosnian samples from Finland and that I might want to compare these with Jabalya. As I opened the box, I first checked whether all the samples that were listed were there and then started to separate the males from the females. I immediately noticed the small number of male samples among them and reasoned that this bias had to do with the origin of the samples.[50] I remarked on this difference to a colleague. Not aware that I was referring to the Bosnian samples, she answered that she was not surprised. "It's always the case. There are always fewer males than females." She stated further that because samples are often collected in collaboration with medical teams, women tend to participate more often than males.

Lab P had tried to tackle the bias towards female samples by asking explicitly for male samples, as in the case of the Dutch samples. The newly delivered samples from the Middle East were all male samples, and when the laboratory asked for the Russian sample they stated clearly their special interest in males.

Changing practices, making sexes

This account about the contribution of cell material by women and men when samples are collected shows a sexual bias in the availability of samples for population studies. The joint work of geneticists and medical teams may clarify the variety in the sampled populations. However, it does not necessarily explain the bias in Lab P collection. So let us take a look at how this bias may be viewed within the context of that laboratory.

As I have pointed out, samples may enter the laboratory without a specific research objective, simply by being offered to the laboratory or because the laboratory members know someone who has rare samples difficult to obtain otherwise. These samples just sit there until a project comes up in which they could become important. Such samples may reflect the specific variety produced in fieldwork. Other samples, however, are requested because there is already a project in which they are deemed significant. Such was the case for the largest part of the samples that arrived when I was in Lab P. In almost all cases, these were male samples. This indicates that under specific conditions it was possible for the laboratory to alter the sampling bias produced in fieldwork. Consequently, the ratio of male and female samples in Lab P does not necessarily have to represent the ratio that occurs in fieldwork where samples are collected. After all, the virtual absence of European samples in the laboratory cannot be said to represent fieldwork either, if one considers the long-standing history of medical studies that have made available large amounts of human blood or tissue.[51]

This shows that the lack of male samples had not previously been perceived as a problem.[52] It hints at a practice in which the sexes were not actively being performed and the samples did not differ in terms of sex. The scarcity in males became apparent when there was a change from a practice where there was no sex to one where the sexes are performed as males and non-males. To relate this back to the traffic in samples as gifts between laboratories and scientific groups, one could say that the sex of the gift emerges in scientific practices in which the sexes are made relevant. The gift of samples is not male or female by nature, but it acquires this quality through ongoing projects and the technologies applied in the laboratory.[53]

In practices of genealogy, the sexes were enacted as patterns of inheritance. A DNA practice, however, showed a more complex picture. The sexes were either absent, as in the case of mtDNA and in the spatial organization of the samples,[54] or they were enacted as a variety of things, as was the case for Y-chromosomal DNA. What sex *is* appeared to be a product of interfering practices such as written records, anatomy, social roles, quality of a DNA sample and the laboratory practice where specific marker technology is being applied. In the final part of this chapter, we will examine a third site, that of genetic lineage, and view how sex matters in a practice where DNA is treated as a technology to produce accounts about lineage and the history of populations.

Genealogy: technologies of lineage/technology of DNA

As has been shown above, there are differences between mtDNA and Y-chromosomal DNA. These differences involve not only their pattern of inheritance or the way they contribute to the production of sexualized lineages or sexualized samples but also their locus in the cell. Whereas the Y-chromosome can be found in the nucleus, the mtDNA is located in the cytoplasm that surrounds the nucleus. But there are more. The number of Y-chromosomes and mtDNA molecules differ per human cell. While the human cell, if from a male, carries one Y-chromosome, it may contain up to a 100 000 copies of mtDNAs. This is the reason why, compared with Y-chromosomal DNA, mtDNA is convenient to work with using old samples. The two also differ in shape and size: mtDNA is a circular genome consisting of 16 500 base-pairs whereas the Y-chromosome is linear and consists of 60 million base-pairs. More important for studies of genetic lineage, they differ also in their mutation rate. The non-coding part of mtDNA, and this is the region of interest in these studies, mutates 20 times faster than the non-coding part of the Y-chromosome. This results in a much higher diversity in mtDNA.[55] As stated above, mutations are read as historical events in studies of genetic lineages. They contain information not only about genetic lineage but also about events in the history of humans or a population.

I will explore this through the example of scientific papers. Two publications of Lab P comparing both systems will be discussed. These papers study the above-mentioned "bottlenecks" in Finnish and Sinai populations. By so doing we will see how in studies of genetic lineage DNA is handled not only as a (standardized) technology to produce sexualized lineages but also as yet another enactment of sex difference.

The papers reported a reduction in Y-chromosomal diversity in the Finns and in two populations in El Sawarka and El Bayadia (north of the Sinai).[56] The mtDNA data, however, was found to be just as diverse as in the surrounding populations of both Finland and the Sinai. The reduced diversity on the Y-chromosome suggested a bottleneck, caused either by a reduction in population size or by a founder effect (a small number of individuals contributed to the contemporary genetic variation). The question, of course, was how to understand these differences in diversity. The reproductive success of some Finnish males over others, in the sense that a small number of males had succeeded in passing on their genetic material whereas others had failed to do so, was pondered as a possible explanation for this phenomena in the Finnish study. Alternatively, one could conclude that the "colonization" of Finland was the work of a great number of women and only a small number of men. In the case of the Sinai, the results seemed to confirm notions about marriage patterns in the populations studied. Male polygamy and cousin marriages are frequent in this part of the world and women marrying outside would leave their groups to live in that of their husbands. Hence "the traffic in women" is here suggested to be the main source of diversity in the Sinai populations.[57] The "lack of traffic in males" was considered the reason for the low Y-chromosomal diversity.

To analyze the difference between the high diversity on the mtDNA and the low diversity on the Y-chromosome further, the difference in mutation rate was considered by both papers. It was argued that the non-coding mtDNA, also called the control region, is not only known to mutate 20 times faster than the non-coding Y-chromosomal DNA, but that this rate also varies within the control region and that certain positions within this region would mutate up to 15 times faster than others.[58] Therefore, in comparison with Y-chromosomal non-coding DNA, some positions of the control region may mutate 35 times faster. The control region would consist of "slowly evolving (nucleotide) positions" and "fast evolving (nucleotide) positions." Variation in the first would indicate events that had occurred earlier in time and the variation in the latter was considered to be more recent. Taking this information into account and by focusing on slowly evolving positions, it was found that the mtDNA diversity in Finns would be reduced to match that found in Y-chromosomal DNA,[59] whereas the mtDNA diversity in the Sinai would remain higher than that found on the Y-chromosome, and just as high as the diversity found in surrounding populations (along the Nile and in the Nile Valley). As a result, it was concluded that the bottleneck in the Finnish population was a reduction in population size and that it had affected males as well as females. For the Sinai, however, the reduced diversity on the Y-chromosome could be viewed as a founder effect and confirmed that marriage patterns (traffic in women and lack of that in males) explained the discrepancy between the two diversities. Moreover, in the Finnish case, the diversity found in rapidly mutating positions of the control region was used to date the bottleneck. By assuming that those mutations occurred after the bottleneck event and that such mutations take place once in a 1000 years, the bottleneck was estimated to have occurred 3900 years ago. The same procedure suggested that in the Sinai "the patterns of marriage must have been upheld over substantial time"[60] and that future research should make it possible to study the age of these patterns.

Doing lineage: making sexes

I have argued above that mtDNA and Y-chromosomal DNA provide geneticists with a two-sexed model for studying genetic lineage. The Sinai and Finnish papers showed how DNA can be handled as both data and technology to do exactly that. They also showed more. However, let us first take a look at how DNA is treated as a technology.

Putting genetic data in the context of lineage is not straightforward. Because they lack recombination, both the Y-chromosome and mtDNA are viewed as molecular clocks.[61] This notion implies that their mutations are assumed to occur at the same rate in each individual. Consequently, the number of mutations in each of these systems is assumed to correlate with a historical time. The more mutations, the more time they took to occur. Determining the time between two mutations is a crucial part of establishing genetic lineage. As stated earlier, mtDNA, in general, mutates at a much higher rate than the Y-chromosome. The mtDNA clock ticks faster than the Y-chromosomal clock. Yet both types of information are treated as complementary, indicating a parallel historical event. Consequently, these clocks have to be calibrated in order to discern a comparable historical time. To do this, it is not enough to know about patterns of inheritance: that the Y-chromosome shows the paternal line of inheritance and the mtDNA the maternal line. Specifically, the Finnish case showed that the high diversity in mtDNAs in the Finns was not taken at "face value." The control region of the mtDNA was treated as a system consisting of different clocks indicating different time calendars. By taking into account only the slowly evolving sites of the control region (as historical records), both that information and that for the Y-chromosome became standardized and compatible, contributing to an account of an historical event. When the event was identified as one affecting males and females equally, the information found in the rapidly evolving sites of the control region could be applied to date that event as a point in history. So the time on the clocks had to be discerned. By presupposing the time between two mutations in the rapidly evolving sites, the time of the events of interest could be set on the complementary clocks of both mtDNA and Y-chromosome. This shows that the mtDNA system is handled as a technology contributing to accounts produced about the past of a population. Moreover, mtDNA is handled as both a technology (a calibrating device for setting molecular clocks) and a resource containing information comparable to that found on the Y-chromosome.[62]

However DNA is not alone in producing accounts. From the reduction in diversity in the Y-chromosomes of the Finns, it was suggested that a small number of males contributed to contemporary genetic variation. This bias in the number of males and females was the very reason for considering the slowly evolving sites of the mtDNA.[63] It indicated that the Y-chromosome came to stand for men and the mtDNA for women, both living in the same period of time. Therefore, although the two systems of inheritance point back to *partial ancestors* that do not necessarily have to coincide with individuals or parents (MRCAs), the presupposition of individuals and parents is necessary for studies of genetic lineage. The Finnish Y-chromosomal data was read as *men* and the reduction in diversity as a reduction in the number of men, raising the question of whether the same was the case for *women*. Hence further analyses of mtDNAs. Similarly, the Sinai reduction in diversity on the Y-chromosome was placed in the context of social relations of marriage patterns, which should explain the lower diversity in men compared with women. Hence, in studies of genetic lineage, sex is not only performed as a pattern of inheritance. To be able to make sense of various types of information based on mtDNA and Y-chromosomal data and to produce an account of human history, these particles have to stand for woman and man who pass on their genetic material in a socio-historical context.

Despite all the differences between mtDNA and Y-chromosome, in the context of genetic lineage, the two systems are considered *comparable* because both are passed on without recombination. They are also deemed *compatible* because they point back to human ancestors in both mtDNA and Y-chromosome who are supposed to have lived at the same historical time. They are also considered to be *complementary* because they tell two parallel human histories, that of men and women. They are considered to be molecular clocks, each ticking at a different speed. However, from the Sinai–Finnish example, it became clear that these clocks can be set in such ways as to give the same time whenever you look back into history, revealing men and women.

Conclusions

In genetics, XX and XY are just two of the various way of knowing the sexes. Throughout this chapter, I have located the sexes in practices of genetic lineage, I have traced the relevance and irrelevance of the sexes and have examined how the sexes are enacted in such practices. The analyses make clear that genetic sex is a *doing*, and that the various ways of performing this, the various ways of doing genetic sex in practices, goes well beyond an identity that can be located in the individual – or in the DNA for that matter. Sex difference is just one form of doing genetic lineage, and in itself it consists of many things. To be sure, genetic sex is not a list of references to an individual. Such a list would rather point to practices of doing genetic lineage. Therefore, to study the sexes and the differences between them is to study the practices in which they are performed.

In studies of genetic lineage, geneticists aim at giving an account of human history based on the DNA. In these studies, DNA is treated both as a resource for learning about similarities and differences and as a technology to establish lineage. I have shown that in a practice of mtDNA and of Y-chromosomal DNA, the handling of DNA as a technology helps to establish sexualized lineages. This, in its turn, affects the treatment of DNA as resource. DNA samples, as in the case of Lab P, were no longer just population samples but were enacted as male and female samples. Moreover, given the aim of studying the history of populations, the handling of DNA as a technology assisted the essentialization of sexual differences. Differences and similarities in the DNA could thus be read back onto the history of populations, producing men and women. This indicates that studying the history of humans via the DNA subsumes the diversity in practices of doing genetic lineage and the various ways of performing the sexes in laboratories.

Might this then lead to the conclusion that, in the end, genetics does the same old thing: that it makes "biological" categories, and that feminists should keep an eye on how men and women are done outside this field? The aim of this chapter, however, was to show that genetic sex and sexual differences do not exist by themselves but are enabled by technologies. These very technologies affect not only our ways of knowing genealogy, lineage, parenthood and individuality but also the practice of genetics itself. Here I have shown that the technical possibility of studying Y-chromosomal DNA has introduced a sexualized system in the laboratory. Technologies thus affect what the sexes are made to be. The *heterogeneity* of scientific practice examined here may give hints about how to do biology according to the *nature* and heterogeneity of scientific practice.

Notes

1. In colloquial speech between geneticists the laboratories are usually referred to as the Pääbo laboratory and the forensic laboratory, hence my choice to refer to them as Lab P and Lab F.
2. Linkage disequilibrium is the non-random association of alleles on chromosomes: i.e. the phenomenon that different alleles or genes are linked in their pattern of inheritance. A popular example of this is that of hair and eye color and the specific combinations in which they are inherited. It should be noted that Maris Laan's project did not target such phenotypic traits but rather what she called "anonymous loci" and their frequencies in different populations (Laan and Pääbo, 1997). Note that the X-chromosome does not escape recombination in general, because it may just as well stem from a female individual, but Maris Laan looked at markers in the isomers where the chances of recombination are the lowest.
3. For registrations of viruses and their translations once they start to travel, see de Bont (2000).
4. See Hirschauer and Mol (1995).
5. Both feminism and feminist studies of the sciences constitute a large terrain. I can hardly do justice to the diversity within these domains nor acknowledge the

inspiring work of different kinds, such as that of Evelyn Fox Keller, Sally Hacker, Donna Haraway, Judith Butler and Annemarie Mol, by talking simply, as if it were that, about feminist scholars. Nor can I do justice to the elaborate studies contributed to the field of science studies by feminist scholars and feminists in general. For early contributions to feminist studies of the sciences see, for examples, Harding (1986); Bleier (1988). For an insightful study on feminisms (liberal, Marxist, radical and socialist feminism), see Jaggar (1983). My interest in genetics is very much indebted to and inspired by the work of Evelyn Fox Keller (1985, 1992a,b, 1995a,b, 1998). Sally Hacker (1989) taught me that one should engage in science and technology to make a political difference, Donna Haraway's (1991b, 1992) approach gave me the promise of combining socialist, feminist and anti-racist politics with academic work and doing science studies. The work of Annemarie Mol and Judith Butler (1990) showed that there were other ways of theorizing sex or gender, making it possible for me to relate to it. Moreover, from Annemarie Mol's work, I learned to focus on processes of doing science and their normative involvement in scientific objects (Mol, 1985; Hirschauer and Mol, 1995). On sex and race I have benefited from the work of Stepan (1986); Mol (1991a); Haraway (1992); and many others.

6. My use of "sex" and not "gender" is motivated, even though the English language puts some constraints on its use. I apply it not only to destabilize the seemingly neat distinction between sex (as being biology) and gender (as being culture), but also in accord with my claim that culture is part and parcel of genetic knowledge. In her study on primatology, Donna Haraway (1988: 95) stated the following with regard to the distinction between sex and gender: "The boundary between sex and gender is the boundary between animal and human, a very potent optical illusion and technical achievement." For an elaboration on sex and gender and especially on its use in Dutch feminist studies, see Mol (1998). On the traffic of gender, as a category/term, between languages and on distinctions between sex and gender, see Haraway (1991c).

7. In the case she discusses, Annemarie Mol argues that atherosclerosis means different things and she shows various ways in which it is performed in different sites: in the general practitioner's surgery or in statistical records, in the Netherlands or in the tropics, in the body or in technologies. This chapter is in various ways akin to the work of Annemarie Mol, especially to the method she proposes (Mol, 1985, 1990, 1991a, 2002; Mol and Hendriks, 1995).

8. Keller's (1986) elegant paper on the issue of gender and science tried to go beyond, or to find diversity between, the feminine–masculine dichotomy. It should be noted that Keller did not elaborate on this aspect of counting, namely that there might as well not be anything there to count. In her paper, Keller engaged in a discussion about biological discourse, but along with many feminist scholars she did not address the issue of biological sex. The latter is taken as the unproblematic and not so relevant fact in the complex configuration of cultural sex determination, namely gender. For a dicussion of early feminist work in relation to biological sex, see Haraway (1991d) For work that does take the matter of biology seriously, see the work of Donna Haraway and that of Annemarie Mol (notes 5 and 6); on the materiality of sex difference, see Oudshoorn (1994); for a critique of the stability of biological sex, see Butler (1993); Hirschauer (1998).

9. Judith Butler (1993: 9) proposed the notion of "matter" as an alternative to notions of construction. I follow this notion, which she defined as "*a process of materialization that stabilizes over time to produce the effect of boundaries, fixing, and surface we call matter*" (italic in original). The notion of matter is akin to that

of performance or enactment, which I am more inclined to use, specifically because the latter focuses more on the process of performing an object rather than the end-product. Moreover, I hesitate to apply the term matter because of the understanding it conveys. Although for Butler it means both, matter is commonly understood as "to matter," to be important, rather than "to become matter," to become material. Haraway would call this effect crusted language, see M'charek and Rommes (1998). On performativity, see note 73 in Chapter 1.

10. In this chapter, I use performance and enactment interchangeably; see note 73 in Chapter 1. On the performativity of sex, see Hirschauer and Mol (1995), Butler (1990, 1993), Hirschauer (1998); on (inter)sex(uality) as an achievement and passing (i.e. an active performance and a management of complex rules, rights and regulation to live in the elected sex status), see Garfinkel (1996a).

11. Emma Goldman's (1970) political essay on women and prostitution was first published towards the beginning of the twentieth century in the periodical *Mother Earth*; Rubin's (1975) essay on gender and kinship was published in the mid-1970s. Whereas Goldman viewed sex as a commodity in an economy of exchange, namely prostitution, where men and women are assigned agency, Rubin focused on the exchange in kinship relations, where men have more rights over women because of the nature of the exchange, namely an exchange between men of the "object" women, the latter being deprived of those or equal rights and selfhood. For an elegant comment on Rubin, especially on the issue of agency in exchange relations, see Strathern (1988: 328–334).

12. The various projects conducted in Lab P were subdivided into five groups. For example, the different projects concerned with human population histories were part of the Population Group. All groups met on a weekly basis with the professor to discuss the individual projects and how they were progressing.

13. This large amount of samples was, in fact, available as blood samples in Lab P. However, as the human immunodeficiency virus (HIV) research group of Professor Jaap Goudsmit (Academic Medical Centre, Department of Human Retrovirology) in Amsterdam became interested in these samples for family studies, and since the samples consisted of so-called trios (father, mother and child), the HIV laboratory offered to do the extraction and share the DNA with Lab P, which saved an enormous amount of work.

14. On spatiality, organization of laboratory work and the carrying out of experiments, see Lynch *et al.* (1983), who discussed a study of Friedrich Schrecker. In 1980, he had conducted a laboratory study in a chemistry laboratory/class where students carried out their experiments. Schrecker assisted a disabled student (partially paralysed) in this class. This study showed that laboratory work is embodied and circumstantially contingent, and that experiments are conducted "in the spatiality and arrangement of the laboratory-table's display."

15. The importance of this gift economy becomes apparent at conference meetings. Especially during coffee or lunch breaks, or during dinners, scientists would establish collaborations of various kinds. Furthermore, the acknowledgement section in published papers could also be read as a tribute to the various gifts, in the form of samples or other material, comment and feedback, statistical analyses and various other forms of assistance.

16. For different versions of population in laboratory practice, including Dutchness, see Chapter 2.

17. The exchange economy between Lab P and Lab F was initiated during the conference *Genetic Variation Europe: Genetic Markers* (Barcelona November 1995). Peter de Knijff, the head of Lab F, and I attended this conference and met

The traffic in males

Svante Pääbo, the head of Lab P, and Antti Sajantila. There, collaborations on the Y-chromosome as well as on my own research project were discussed and established; for elaboration on collaborations amongst geneticists, specifically with reference to Y-chromosomal markers, see Chapter 3.

18. von Haeseler *et al.* (1996: 135).
19. von Haeseler *et al.* (1996: 137).
20. The reference to archeology highlights a heated debate between disciplines, namely between genetics and paleontology. The battle is about human origin, and about which discipline has the best access to it. An example of such a debate is that between the paleontologists Alan Thorne and Milford Wolpoff, on the one hand, and Alan Wilson and Rebecca Cann on the other. The issue is not only which sources provide the best entry into human history and origin but also how the spread of humans around the world came about: the multiregional theory versus the African origin theory (Thorne and Wolpoff, 1992; Wilson and Cann, 1992). Also, Mark Stoneking, a population geneticist, talked about this ongoing debate and made clear the privileged view of genetics as follows: "We geneticists know that our genes must have had ancestors, but paleontologists can only hope that their fossils had descendants." (Interview with the author at the Laboratory for Evolution and Human Genetics in Munich, March 11, 1997).
21. Note that I do not wish to discuss different ideas about parenthood or practices of parenthood as a fact of nature, but rather how notions of parenthood contribute to versions of the sexes in studies of genetic lineage. Anthropologists of different kinds have addressed the issue of kinship and reproduction. Traditionally, anthropologists have taken kinship to be that which comes "after the fact of nature." "It is important to realize at the outset that, while the biologist studies kinship in the physical sense, for the social anthropologists kinship is not biology, but a particular social or cultural interpretation of the biological universals just mentioned" (Parkin 1997: 3). Marilyn Strathern (1992, 1995a) was more attune with the approach on kinship I wish to explore here and looked at similarities and dissimilarities between nature and culture. She took kinship as a hybrid connecting these technically assisted domains and as a product of individuality and diversity.
22. See also Strathern (1992: 83–84).
23. For an elaborate examination of different types of genealogical tree, see Bouquet (1995); see also the treatment of genealogical trees in medicines as technologies by Claudia Castañeda (1998).
24. In genetics, genetic lineage in a population refers to the frequencies and to the distributions of variation (in alleles), where a Gaussian curve indicates lineage and a skewed curve indicates the absence of lineage. It goes beyond the scope of my argument to address the statistical technicalities of genetic lineage.
25. See Chapter 4 for a discussion of the controversy over bi-parental inheritance of mtDNA.
26. Similarly Annemarie Mol (1985: 20) argued that certain strands of the life sciences do not sex the individual body but its characteristics. The sexed body emerges, for example in anatomy, only in relation to other bodies assisted by statistics, as being feminine or masculine.
27. Note that another category of MRCA is not considered, namely the X-chromosomal MRCA. The X-chromosome is not an autosomal but a sex chromosome, and it is bi-parentally inherited in females in contrast to its pattern of inheritance in males.
28. Geneticists would always point out that these estimates are rough. Modern humans are estimated to have emerged in Africa about 100 000–200 000 years ago. Whereas this dating would coincide with estimates for mtDNA, and this is not

surprising because the emergence of modern humans was based on mtDNA data, the dating of Y-chromosomal DNA seems to be a little older, namely 270 000 years ago; for mtDNA estimations, see von Haeseler *et al.* (1996: 135); for both estimates, see Pääbo (1995); and for Y-chromosomal DNA estimates, see Dorit *et al.* (1995).

29. For an analysis of how both DNA and the cell are treated as technologies, see Rheinberger (1997: 275). Who stated: "Diese Enzyme und sonstigen gereinigten Moleküle stellen eine Art 'weicher' Technologie dar, eine molekulare Technologie, die der Lebensprozeß selbst über eine Periode von Milliarden Jahren entwickelt hat." See also Rheinberger (2001).

30. See Sajantila *et al.* (1996).

31. Professor Bonne-Tamir visited Lab P when I was there. She gave a seminar in which she addressed the issue of sampling, how it was conducted in 1967 and showed slides of some of the populations as well as the scientists.

32. Blood serum is the fluid that remains when blood has clotted.

33. Primers are necessary for copying or cloning of DNA fragments using PCR. They make up the beginning and end of such a fragment and are synthetically produced. The ladder could be seen as an extra sample, which works as reference because it contains all possible variations that could occur in such a DNA fragment and, therefore, helps to estimate the size of the fragments in the samples under study. Some ladders are called universal. They do not contain the fragment sizes that can be found for a specific marker, but standard ones such as 50, 100, 150 base-pairs, etc. These ladders, also called sizers, serve as molecular weight markers (Hartl, 1995: 379). See also Chapter 3, where I describe PCR and the components necessary for it more extensively.

34. For allele, see Chapter 3.

35. The origin story of Jabalya was that it was founded in the seventeenth century as a monastery by Christian monks. Its population became intermixed as a result of both pilgrimages and the fact that it was traditionally a passage point for those wishing to cross the desert.

36. See note 6 in Chapter 2.

37. See note 16 in Chapter 3.

38. Again I would like to emphasize that the distinction between doing genetic lineage and doing DNA is not ontological. Genetic lineage is involved in daily laboratory work, hence my move from Sawarka to Jabalya. In general, analyzing the data may include doing more DNA, looking at more samples, consulting the data of colleague laboratories, etc.

39. Anderson *et al.* (1981); Wilson *et al.* (1985); Cann *et al.* (1987).

40. Roewer *et al.* (1992); Dorit *et al.* (1995).

41. The viability of these markers was acknowledged in a round-table discussion in 1995 at the conference *Human Genome Variation in Europe: DNA Markers* (Bertranpeteit, 1995): for an elaboration, see Chapter 3.

42. See Sajantila *et al.* (1996); Salem *et al.* (1996).

43. See Garfinkel (1996b; especially Chapters 1 and 7; see also Michael Lynch (1997; especially Chapter 1).

44. Lab P does not assign new numbers to samples that come into the laboratory. Whatever number the samples have, this is how they enter the records. It is a practical method because most of the samples are collected by other laboratories or scientific groups. Keeping records of who supplied the samples and when, and adopting their nomenclature, offers a way of communicating what is and is not already in the laboratory when new deliveries come in.

45. For such interferences, see Mol (1985).

46. Interference is a term of Donna Haraway (1991a). For an elaboration on interfering practices, see Law (2000). On the relevance of breasts in medical practices of trans-sexuality, see Hirschauer (1998).
47. Old samples work better for mtDNA than for nuclear DNA. The reason for this is that each human cell may contain up to 100 000 copies of the same mtDNA and only one copy of nuclear DNA, the latter being divided over the 46 chromosomes. If DNA starts to deteriorate, there is a fair chance that there will still be some mtDNA molecules remaining that are useful for study, which cannot always be said for nuclear DNA.
48. Professor Svante Pääbo, interview with the author at the Laboratory for Evolution and Human Genetics in Munich, 4th February 1997.
49. On lack of stability of a person's sex and the need for it to be actively performed to keep it, see Hirschauer and Mol (1995). For a similar case in medical practice, see Mol (2000, 2002).
50. The Bosnian samples were collected for a forensic study attempting to identify the victims of the war massacre so came from the corpses found in the mass graves. It was not clear whether the samples we received in the laboratory were from the dead victims or their relatives.
51. On the gifting of blood to blood banks in a number of countries, especially in the UK, and the USA, see Titmuss (1997); on medical trials, see Kevles (1985).
52. For example, the geneticists Mark Jobling and Chris Tyler-Smith (1995: 455), both conducting work on the Y-chromosomes, urged the geneticist engaged in collecting samples within the context of the Diversity Project to take notice of this bias and to include adequate numbers of male samples.
53. See Strathern (1988).
54. The absence of the sexes is not exceptional. This would also hold for the twenty-two autosomal (non-sex) chromosomes. Since all human individuals carry a set of two for each of these chromosomes, sex does not matter here either.
55. This is one of the reasons why the Y-chromosome had been insufficiently studied. It proved to be difficult to find polymorphisms on the Y-chromosomes and those that were found were not deemed suitable for studies of lineage and evolution. For a good and comprehensive review of Y-chromosomal studies; see Jobling and Tyler-Smith (1995).
56. Sajantila *et al.* (1996); Salem *et al.* (1996).
57. In this respect, Luca Cavalli-Sforza stated the following in a lecture: "the Y-chromosome is highly geographically clustered, compared with mtDNA, which is far less clustered. There are two reasons for this. One is that there are so many mutations on mtDNA and two that women tend to migrate more than men." He added jokingly: "*La Donna e Mobile*," referring to the opera of Giuseppe Verdi, *Rigoletto*. The lecture was given at the *Fifth Annual Meeting of the Society for Molecular Biology and Evolution*, Garmisch-Partenkirchen, Germany, June 1997.
58. This phenomenon is also known to be the case for Y-chromosomal non-coding DNA but is not addressed in these publications of Lab P. On various different mutation rates for the Y-chromosome, see Jobling and Tyler-Smith (1995: 450). For different mutation rates in the control region (mtDNA), see Hasegawa *et al.* (1993: 350); Wakely (1993).
59. This is the case if only positions that change once in 13 000 years are considered, (Sajantila *et al.*, 1996).
60. Salem *et al.* (1996: 742).

61. See also Chapter 4.
62. This hybrid feature of mtDNA, being technology and resource, is not exceptional. I have shown this to be the case for genetic markers and for the Anderson sequence, see Chapters 3 and 4.
63. Svante Pääbo, personal communication, Munich 1997.

6

Technologies of similarities and differences, or how to do politics with DNA

This book has dealt with the socio-materiality of genetic diversity in the era of the Diversity Project. It started off by asking what population is, did the same for technology, then went on to consider the autonomous nature of technology and finally that of DNA. Four cases and a few more practices have been examined. The localities investigated were laboratories. Both scientists and DNA were deliberately kept out of focus. Many other aspects were placed center stage. Technologies, individuals, populations, lineage, gifts, races, sexes and blood – among others things – have been central in the previous chapters. They were made part and parcel of the laboratories studied. The analyses examined the heterogeneity of technologies and practices and how these affect the object of geneticists' research. Now it is time to make some links that go beyond each individual case and to narrate stories that reach beyond the Diversity Project. I would like to tell three stories, each aimed at embedding the results of my research in other academic fields: science and technology studies (STS), population genetics as related to the Diversity Project, and gender and anti-racist studies. After each story, I will make a number of points.

I will conclude this chapter by some notes on method.

Naturalization: tracing the politics of nature and technology

Story 1: talking back to STS

This book is a study of laboratories. Such studies are not new in STS. Laboratory studies have been part of the scene ever since the late 1970s. Having studied the Diversity Project in laboratories, I would not want to argue that these are privileged sites for studying genetic diversity. Laboratories were rather chosen as a point of contrast to the global discourse of the project.[1] Whereas the goal

148

of the Diversity Project is to map what is out there, namely the diversity of the world's populations, my aim was to investigate the locally achieved character of genetic diversity in the process of laboratory conduct. Instead of treating genetic diversity as something that is inherent in individuals, populations or their DNA, I have traced it in technologies and practices. These localities did, indeed, allow me to show that neither DNA nor scientists work by themselves. In addition, the localities showed that laboratories are no isolates, producing exotic kinds of knowledge, nor are they easily *taken up* in a global endeavor to become dislocated spaces. Laboratories do not meet the classic distinction between local and global.[2] Both are performed simultaneously. Laboratories are best seen as sites of interference, of nature and technologies of various kinds. Consequently rather than a unified and a well-ordered process, the work of laboratories can be understood as the management of contingencies.[3] The heterogeneity of such practices is in itself enabling. It allows for the making of new links, the solving of practical problems and the establishing of lineages between labs.

So far, so good. At least for an STS audience there is nothing strange going on here. However, my goal was not to map out the work of the laboratories or the organization of science in the Diversity Project as such. My interest lay in analyzing the very *topic* of the project: genetic diversity. More specifically, my interest is in issues of normativity and politics. How is genetic diversity practiced? What have we learned about it from the daily work in the laboratory? And what kinds of story did the analyses in the different chapters together produce about that work? One of the themes narrated in this book is that of *naturalization*. Let us take it up again here and draw together some points about naturalization.

Is there such a thing as a natural object, for instance population?

Population is not just any category. It is crucial to *population* genetics, its major object of study. In Chapter 1 we saw that population was a matter of debate, also within the confines of the Diversity Project. I examined what population "is." The first thing I noted was how geneticists in the Diversity Project *defined* it, namely on the basis of linguistic separations. The second was how it is practiced in laboratories. That was something I examined specifically. In analyzing a forensic case, I showed that population is not one unified category. In this particular case, at least seven different versions of populations were practiced. Each version of population was enabled by different technologies and different practices. This does not, however, imply that different technologies produce different insights about a pre-existing object. Population is neither a matter of nature, one that can be discovered, nor a matter of definition, which represents a singular object out there. My point is rather that population is enacted with the very technologies aimed at studying it. Moreover, the different versions of population do not add up to produce a better "re-presentation" of it. Some versions might conflict. For example, enacting population as national identity does not map onto one enacted as racial differences; population differences on the bases of mtDNA do not necessarily coincide with those based on nuclear DNA. Population is manifold and this very fact underscores the heterogeneity of laboratory practice. Different practices produce different versions of population.[4]

The question prompted by this is how is it that there exists such a thing as population, or even a field called population genetics. Why doesn't population fall apart or cease to exist as an object? Annemarie Mol and John Law have suggested coordination as a way of making a manifolded object pass as one. Coordination can be achieved in various ways.[5] One way of doing that is by shifting focus, while moving from the one practice to the other for the purpose of comparison. For example, a collection of 153 samples taken from 35-year-old males living in the Dutch town called Doetinchem and collected in a large-scale study on heart disease entered the laboratory records as the Dutch samples. As such (a collection of samples plus their enactment in the records as the Dutch population) they are ready to be compared with, say, the German population (however assembled). Two populations *compared*; two populations *translated* into one another: translated into one another because, as Strathern argued "comparability is not intrinsic" to the objects compared.[6] Rather it is the act of comparing that produces connections and relations of similarities and differences. So the practice of comparing is one way of coordinating different versions of population, of making population more stable and to pass as one. Another form of coordination is established by shifting focus away from population towards genetic lineage: by making different versions of population *compatible*. Compatibility as a practice of connecting and making disparate objects converse with each other is established by placing population in the context of origin and lineage. Departing from the presupposition that all humans have one single origin, the history of a population, that is when and how members migrated from one place to the other and the age of a population, can be traced through studies of lineage. Even though genetic lineage and population come as a pair, the focus in such studies is more on the former than on the latter. In this practice, different versions of a population, whether enacted as race difference, national identity, linguistic characteristics or as geographical location, can be made *compatible*: for example, when combining data about a specific population that were produced in different studies and different places and time. They are taken to refer to a natural process out there, which can be understood through studies of origin and lineage. In laboratory practice, population is enacted as various things that, through the work of coordination, may be drawn together to create the "illusion" of a coherent object.

Laboratories are populated by technologies, but what is technology?

Questioning what technology comprises is hardly remarkable, for many scholars in the field of STS have raised it. Whether the issue was electric vehicles, airplanes, subways, or photoelectric lighting kits, the question has been addressed. In these and other examples it was shown that technologies are best understood as networks in which human and non-human actors are enrolled and translated in order to keep a technology together.[7] My interest in posing the question again has to do with a specific phenomenon of the life sciences at large and genetics in specific, namely the double quality of objects and technologies. In the laboratory, a DNA fragment may be performed as an object of study, but also as a technology.[8] What does it take to make a "biological" object into a technology? To answer the question, I focused on genetic markers and examined how they were applied in laboratory practice. In DNA-based genetics, markers are the tools of comparisons between individuals or populations. They allow for comparisons based on differences. In such a practice, people are not so much related by blood or by DNA but by genetic markers. But what is a marker? The definitions describe markers as DNA fragments containing specific information; a marker is thus an *object* of comparison. In the laboratory, however, a marker is performed as a *visualizing technology*, which is, among other things, dependent on protocols, chemicals, PCR programs to copy a specific DNA fragment, particularities of such a fragment. The successful alignment of all these components for the purpose of visualization contributes to or inhibits possibilities of applying a marker as an object of comparison. Yet, there is more to markers. Good genetic markers should contribute to the analyses of what they reveal. The analyses of data are assisted by statistical models and not just any fragment of DNA can become a good genetic marker for that purpose. The variability (or polymorphism) of a DNA fragment and the specific distribution of alleles in a population make some markers good for a particular study and others not.[9] Methodological analysis is built into good genetic markers from the start. Consequently, next to being an object and a technology, a marker is also a *methodological tool*.

Moreover, my analyses have suggested that a marker is best understood as a local mediation of scientific practice in which humans, technical devices, chemicals, theories, texts and DNA are aligned to achieve a specific goal. The significance and the mutual effects of all these components become apparent particularly when markers start to travel between laboratories. Therefore, the answer to what a marker is points in the direction of scientific work and how that is organized. Rather than an entity, a genetic marker is indeed a socio-technical network. This network may become more or less solid, contributing to the standardization of a marker, or it may remain more or less fluid, allowing for its flexibility and alignment with other practices.[10]

Technologies are locally mediated practices, why do they appear autonomous?

The answer to why technologies appear autonomous is related to the issue of naturalization. In this study, I have investigated a standardized technical device, a mtDNA reference sequence, which is also called the Anderson sequence. This technology appears nowadays on a computer screen as a text, but it is also an object, produced in 1981 on the basis of cell material of two human individuals. In its capacity of reference sequence, it is applied as a means of comparing individuals and as the terms in which their mtDNA sequences are expressed. As a *text*, which consists of 16 569 characters/letters representing the nucleotides, it moves smoothly and is virtually available to any laboratory in the world. This form, however, neither explain its success as a standard nor its naturalization. Looking at the kind of work it enables in laboratories, I have shown that its functioning as the standard is dependent on the reification of a specific DNA practice, that of sequencing DNA and comparing the sequences. Therefore, its usefulness as a technical device depends on how DNA is handled in different laboratories. Moreover, in these practices, it is applied as more than a technical device to produce and compare sequences: it is also treated as the source from which any other sequence has evolved. Sequences in DNA are said to contain mutations *from* the Anderson sequence. One could say that it is treated as the reference sequence not only by *convention* but also by *nature*. However, as a sequence it was produced somewhere in particular practices. Going back into this history of Anderson, I have shown the normativity of the practices: the technologies and tissues applied to produce a composite sequence based on the mtDNA of two individuals. The reference sequence did not leave these practices behind, which explains why it had become again an object of study when a number of scientists, who assumed that it carried some mistakes, decided to re-sequence it. Yet in the context of population genetics, it seemed to move smoothly and the practices that helped to produce it went unnoticed. This ongoing success in different laboratories is not only attributable to the organization of DNA research in such practices but also to the theory of DNA inheritance, the theory which treats all individuals as part of one genealogical family. It is with the help of this universal theory that the Anderson sequence occupies the place of a *natural* object and, in a very practical sense, comes to stand for the sequence from which all other sequences evolved. The results acquired through comparisons with Anderson can, with the help of this theory, be read back into nature to reveal the history of a population or genetic lineage. Hence, the practicability of Anderson and its treatment as an autonomous object in the laboratory results not only from how DNA is handled but also from how sequences are analyzed; that is, which theory is being applied to accomplish this. One could say that naturalization of technologies is not just a matter of comparable practices but also one of a unified "world view" in a scientific field.[11]

The traffic in things is common practice in laboratory work and the alignment of humans, objects, technical devices and theories is part of making things work. Where, then, lies the problem of naturalization? This problem is not so much in the fact that technologies are locally produced and that they embody specific practices. Rather the problem lies in the treatment of the results enabled by them. What is naturalized in the case of Anderson is not only the reference sequence itself, as a kind of original sequence, but also all the similarities and the differences produced through comparison with it. In what follows, I will address why this is a problem.

So technologies are locally mediated practices that, with the help of unifying theories, may be treated as "natural"

Might this statement also hold for the objects of genetics, such as population or DNA? After all, both genetic markers and the Anderson sequence are objects as well as technologies. We have touched upon these issues above, but I have also addressed them in my examination of sex differences. Just as for population, in genetic practice there are various different ways of enacting the sexes. The relevance of sex difference and the way the sexes are performed depends on technologies. In the case examined in Chapter 5, the sex of samples did not exist in the laboratory work until the introduction of Y-chromosomal markers. From that moment onward, it became relevant to know which samples were taken from females and which from males. So different practices introduce different ways of enacting the sexes. One could say that in laboratories both sex differences and population are denaturalized and their manifoldedness is appreciated. The open-endedness of research requires that different versions of such objects can be enacted in the process of study. It enables new links, in terms of analyses or research strategies. However, placed in the context of genealogy and lineage, the different versions of the sexes are either subsumed or added up, and data based on sex difference come to stand for people: men, women and populations. This indicates that, in the process of making universal claims about genealogy and lineage, not only technology but also genetic objects are naturalized. Naturalized technologies are indeed a matter of concern because they have the naturalization of objects amd the essentialization of similarities and differences as their effect.

Should this then lead to the conclusion that geneticists in the Diversity Project tell an old story with new means? The answer is probably both yes and no, and this is exactly the trouble with genetics. It depends, for example, on which story is being told: that of genetic lineage and origin or that of populations. Whereas in the first, population is treated as a passage point of specific genetic information (the vehicles for the spread of genes), in the second it is treated as the embodiment of similarities and differences. Ironically enough, population tends to be naturalized in the former and heterogeneously configured in the latter. This has to do with the difference between a focus on genealogy and origin and a focus on population and diversity. I will elaborate on this below.

Standardization: tracing the normativity of practices

Story 2: talking back to genetics

The fate of scientific results is in the hands of their future users.[12] Geneticists will be the first to subscribe to this statement. Ever since the Second World War, population geneticists have become sensitized to the consequences of science. In the face of looming technological potential, questions about the "effects" of genetics are being raised, even in laboratories. These questions may concern not only racial issues but also the possible psychosocial effects of paternity testing or genetic diagnostics.[13] Here I want to pick up on these moral issues not by taking the questions a stage further but rather by setting them back a step in the trajectory of scientific conduct. In fact the question here is: "What would happen if we decided *not* to make a separation between the worlds of producers and those of users?" "What would happen if we took the producers of scientific results to be their very users?"[14] Instead of following the facts outside the laboratories, I turned my attention towards what happens in those environments. It was there that I wanted to learn about the *stuff* from which genetic diversity was made. What this diversity entails and how it is enabled by technology were my leading concerns. What moral issues did my analyses uncover and what can we learn from these in debates about the Diversity Project?

Working together requires common ground. Geneticists are aware of that. They are also aware that the work to achieve common ground is done by both people and technologies. This is even more so if the aim is an international project. Hence the topic of various conferences organized within the Diversity Project in the early 1990s.[15] Geneticists participating in the project agreed on the terms of reference concerning technologies to be used and the concept of population to be applied. The aim was to standardize both genetic markers and population so as to facilitate the exchange of results and their comparability. However, standards are by no means neutral and their effects may go well beyond the convenient.[16] Therefore, the second theme that I want to discuss is about *standardization*. I want to focus on that and to redistribute some of the moral/normative questions that have been raised within the Diversity Project.

Given the aim of making a genetic map of the world, one of the first
issues raised within the Diversity Project was that of population
Geneticists decided to define population on the basis of linguistic separation. However, my analyses of how population was practiced showed that what population was made to be in laboratories, varies. I suggested that the various ways in which population was enacted had advantages in such locales because of the heterogeneous nature of laboratory work. In laboratories, geneticists are not working on mapping the world but on various different and more specific questions. In addition, the organization of science is based on problem-directed collaborations, rather than general aims, a fact which may contribute to how population is practiced. For example, the traffic in samples between laboratories establishes lineages between these laboratories, and these lineages in their own right have implications for which version of population might enter the laboratory.

Does this mean that what population is made to be is a local matter and cannot be standardized? The answer to this depends on where the action is located. If one considers the process of collecting samples or that of typing DNA at the workbench, then the answer is negative.[17] However the various versions of population that may be found there do not exist in isolation. For example, population on the basis of family names may be involved in a collaboration between laboratories; it may thus be found in Leiden, Berlin and Vienna. Standardization is established as a result of the organization of scientific work and is a product of interacting scientific practices. This also indicates that such a standardized approach to population does not necessarily prevent the occurrence of other versions of it.

In addition to this, standardization is also established in how the results leave the laboratories, in the form of evidence in a scientific paper or as information to be stored in the databanks. My examination of scientific papers showed how data are analyzed and how results from different population studies are made comparable and compatible, suggesting that standardizing population is achieved in retrospect. Therefore, standardization may be more a form of coordination of different practices rather than an attempt to make scientists do the same thing in the same way everywhere.[18] Given the aim of the Diversity Project to standardize what population "is," one could say that the process of standardization is not so much achieved through the collection of samples or through the practice of studying these at the bench.[19] Rather it is achieved in the practice of making populations comparable and compatible, whether this is the practice of modeling population data (in papers) or that of data retrieval (in databanks).

If standardizing what population comprises is a matter of interfering practices, how does this affect technology?

The availability of technologies such as PCR and genetic markers is adjudged to smooth the path for diversity research. Markers are a new phenomenon of post-PCR genetics. It is not that they were not there before the 1980s, but they have only become available in large numbers since the advent of PCR, a fact which contributes to the various different ways of studying genetic diversity. Because the number of markers is growing almost daily, geneticists have attempted to compile a list of markers for the Diversity Project. This list prioritizes the use of specific kinds of marker over others and, in fact, aims at standardizing their daily use in laboratories.[20] Taking into account how markers are applied in practice, I showed that they are dependent on the alignment of various constituents. A marker might just as well be a DNA fragment and the variability it might contain, various chemicals and a polymerase enzyme, a copying technology and the PCR program to run it, a visualizing technology, a protocol or a routine way of doing things in the laboratory. Establishing these alignments in one routine practice does not guarantee the success of a marker in other places. The analysis of the use of DNA typing in chimpanzees showed that genetic markers carry such practices along with themselves while traveling between laboratories. Their success depends on the reification of the practices embodied by them. Therefore, the technology for the purpose of diversity studies may be to hand and may be standardized by convention, but to make such a technology work in a local setting requires work, money and time. It is the repetitive success of such collective socio-technical work that makes a technology into a standard.[21]

This is the kind of socio-technical work that biotechnology companies are trying to make less of a burden. In a way, such companies attempt to solidify the socio-technical network that makes up a marker by providing marker *kits* for genetic research. These kits usually consist of a cluster of markers, and the various experimental steps necessary for their visualization are usually reduced in number. They are, therefore, considered time saving. However, complaints frequently heard from practitioners are that commercial kits are very expensive, putting constraints on which laboratories can afford to use them, and the protocols that come with the kits hardly economize on the reagents. Laboratories, therefore, find themselves investing time both in making the kit fit their laboratory conditions and in optimizing and changing the protocols in order to save on the reagents.[22] The example of commercial kits indicates that the locally achieved character of technology and the fact that it does not travel so easily does not mean that its fluidity cannot be "captured" into a more stable form. It does not mean that technology cannot be solidified. Laboratories are over-populated by *frozen moments* of collective socio-technical work, in the form of chemicals, equipment, machines and texts.[23] Yet any kind of technology does not simply co-determine who may or may not become its future users; it also has to be established in a specific local setting.[24] In addition, the various different tasks carried out in a laboratory and their specific configurations of practice also determine the applicability of markers in such a context. For example, studying genetic diversity in a forensic DNA practice or in a combined practice of forensic DNA and paternity testing may co-determine the sets of markers applied in a laboratory. Placing this back in the context of the Diversity Project and its aim of standardization suggests that local settings and laboratory practices for making markers work might put constraints on which markers will become part of its "priority list," and thus on what will become a standard.

The Diversity Project has encountered much criticism and has been accused of being racist

The criticism of inherent racism in the Diversity Project has been countered by leading scientists in the project, who see the project as a potential means of fighting racism by "proving" that there are no such things as biological human races. Moreover, the Diversity Project was initiated as a response to the HGP, which set out to map and sequence one human genome. This sequence genome, based on the DNA of four to five individuals, was deemed to be Eurocentric by population geneticists. They, therefore, suggested a project to map human genetic diversity on a worldwide basis: hence the Diversity Project. In the debate about race and racism in the Diversity Project, on which I elaborated in Chapter 1, my aims have been to take the examinations beyond the discourse of good and bad science or good and bad "genes." Race and racism, in fact, *matter* not only in terms of good or bad intentions but also, and most disquietingly, in various practices, objects and technologies that escape the notice of their daily users and that seem so convenient and benign.[25] Here lies the puzzling matter-reality of race and racism. It implies that the politics of science and of race cannot be seen as something that could be removed surgically to obtain a "neutral" field.

This means that the politics of race should also be examined in the interactions between humans and things and their materialization in practices and techniques where "nothing strange seems to be going on." Throughout this book, I have focused on routines: on how technologies act on practices and help to produce specific versions of objects. Race is no exception in that respect. Any object is enabled, not only by the work of scientists but also by that of technologies. Given these considerations, I find it important to bring technology, and especially routine and standardized technology, into the debates about race and racism.

Standards can never be neutral. They are always produced somewhere in particular kinds of practice and they involve specific social worlds.[26] Yet their functioning as standards and their ability to comply with everyday practice tend to obscure their normative content and make that inaccessible for interrogation. So where, in the words of Susan Leigh Star, does the mess go? Her answer to this question is that: "it doesn't go anywhere; rather, it is the formal representations that attempt to leave it behind".[27] In my analyses of the Anderson sequence, I have tried to draw together some particular "mess" attached to it and situate it back in laboratory practice. Not because I think it should be there all the time, but rather because I contend that there are stakes involved in its accessibility for geneticists and others. As Helen Verran argues: "Being messy and seamy, it acknowledges the actualities of other times and places, and makes the generalizer's accumulation of power more evident – and, for that reason, less certain."[28] In line with these concerns, I have paid attention not only to how Anderson is applied in laboratories nowadays, but also to how it was made. This involved taking into account the technologies and tissues that were necessary for its sequencing in 1981. The first thing I came upon was the fact of it being a composite sequence, based on cell material of two individuals. The second was how it implicated race and race differences. Moreover, while trying to find out about the different sources of DNA used for Anderson, it became clear that race was made relevant in specific practices and not in others. The same tissue used for the reference sequence was racialized in some practices, but not in that of Anderson. This indicates that, while functioning as a standard, a formal text, the normative and racial content of Anderson had become invisible. The sequence passes as an unmarked piece of technology. For example, the heated debates about the HeLa cell line, its origin, its impact on the relatives of Henrietta Lacks (whose cell material were used for the cell line), and the issue of race and racism that it generated had little impact on the community of population geneticists. Even the availability of more publications on the use of the HeLa mtDNA for the production of the Anderson sequence did not generate any debate. The reason for this may be that few geneticists actually knew that HeLa mtDNA was taken up in the Anderson sequence. Once the Anderson sequence was revised in 1999 and HeLa mtDNA replaced by placental mtDNA, population geneticists were committed to use this with neither hesitation nor consideration. My point here is that the endeavor of the Diversity Project is dependent on standardization and a great deal of effort is put into achieving that for technology. Anderson is a case in point. It also provides an example of how standards carry with them a world of difference and similarity making. Even if they appear flat, shiny and smooth, they carry with them a normative or "ideological" content that may affect the world in unexpected ways. This should be a matter of concern for scientists interested in issues of race and racism. It suggests that technologies and routines should be granted more attention than they usually receive in scientific practice.

The statement that "there is no such thing as race," is not enough

The initiators of the Diversity Project want to contribute to the elimination of racism.[29] By producing better knowledge about diversity and lineage their aim is to show that there is no scientific basis for racial differences. As Margaret Lock and others have argued, this a rather naïve notion of the issue.[30] The belief that the facts of science will do the work of anti-racism does not take into account the variety in which race is practiced in society.[31] Nor does it deal with the persistence of arrived biological ideas about, and techniques for enacting, race and other differences.[32] Moreover, it defines the problem of race and racism in terms of knowledge, or rather a knowledge deficit. The notion is that once societies learn about genetics, race and racism will disappear because it is not scientifically sound. However, the Diversity project cannot escape a world where race and racism are realities experienced by the majority of people. Nor can scientific facts by themselves transform practices, especially those practices outside the scope of the project. This indicates that, rather than a universal claim that refers back to the diversity *in* the "genes," we need answers about the diversity *of* "genes," that is about the various different ways in which genes and genetic difference *can* be established. What is needed is a way of increasing awareness of the diversity of scientific practice. Instead of placing the various different techniques in the background and recounting stories about what is found in the DNA, geneticists could make more clear how different technologies contribute again and again to the reassembling of lineage or categories, each time in a different way. In addition, given the heterogeneity in scientific practice, genetics may indeed contribute to the *denaturalization* of differences.

Any object of genetics is manifold. Nevertheless, as I have shown, results of diversity research tend to be naturalized. In that process, the discourse of the Diversity Project oscillates between the practice of *genealogy* and gene pools and that of *population*. Whereas in the first case populations are treated as resource and passage points of genetic information, in the latter case populations are treated as collections of individuals providing insight into the diversity in such groups. I stated above that population tends to be naturalized in the practice of genealogy. How does that take place? In a *practice of genealogy* geneticists aspire to study the migration history of humans and to estimate moments in history when contemporary "populations" diverged or coalesced. The analysis of specific clustering of genetic similarities and differences between "populations" for that purpose requires that geneticists *estimate* the mutation rate in the DNA: how differences have evolved through time. In order to do this, DNA – or, better, specific parts of the DNA – are said to act as molecular clocks. These clocks tick equally fast in all individuals, suggesting that DNA changes equally fast. Specific parts of the DNA are thus treated as standardized technologies that help geneticists to read genetic similarities and differences and establish the history of "populations." How to estimate the mutation rate, that is how to establish the molecular clock, is obviously of major importance for establishing lineage and genealogy. Doing so, geneticists are dependent on pre-existing notions on evolution from fields such as paleontology and evolutionary biology.[33] So, in the *practice of genealogy*, similarities and differences in the DNA become part of a master narrative about human history and population. The very treatment of DNA both as a standardized technology and as a natural resource accounts for the quasi-monitoring of diversity through history and for the naturalization of population and population lineages. In a *practice of population*, however, different DNA systems or technologies are taken into account. Such studies attend to various different ways of clustering population and question it as a homogeneous object. Moreover, such studies tend to pose questions about the past of a population rather than fitting the data into pre-existing accounts of history and lineage. Comparing two population studies, I have shown that a focus on the populations being studied leads scientists to question a standardized mutation rate, as well as pre-given clustering of these populations. Hence in the process of studying various DNA systems in the *practice of population*, the very concept of population is denatured. However, because of the preoccupation of geneticists with human migration history and origin, the heterogeneity tends to be subsumed and population tends to be naturalized and practiced as "race." This should concern geneticists in the Diversity Project, especially those engaged in debates about race and racism. In fact, the standardization of technologies, such as a molecular clock, obscures not only the practices embodied in the technology itself but also the normative content of the objects enabled by it. This is the very reason why stating that there is no such thing as race is not enough. For race is neither fact nor fiction, but rather a matter of doing.

Diversity: the nice thing about DNA is that everybody has it

The heading to this section reflects a comment by Marilyn Strathern.[34]

Story 3: talking back to gender and anti-racist studies

The statement about DNA, namely that "the nice thing about it is that everybody has it" may bring about both egalitarianism and diversity: egalitarianism by making a universal claim about DNA – we are all equal because we all have DNA – and diversity in the sense of making everybody specific – we are all different in our DNA. Both interpretations, however, point to something essentially there in the DNA, and that is not my purpose in using this statement. The reason I introduce it here is to argue that DNA is neither a commodity that can be appropriated or expropriated nor a fixed measure on which to base similarities and differences. Moreover, I have introduced this statement to challenge a tendency within gender and anti-racist studies to side with "culture" rather than "nature." An emphasis on genes and DNA is thus viewed with mistrust. Throughout this book, I have argued that neither nature nor DNA are ever by themselves, and I have shown that culture is part and parcel of genetic differences and similarities. Once more I want to knit the stories together and explore how we might think in a different way about the statement introduced above.

Egalitarianism and diversity are crucial categories in gender and anti-racist studies, as in feminist and anti-racist movements.[35] In brief, one could say that the history of these academic fields and social movements has shown a change of focus during the twentieth century, from a politics of egalitarianism to one of diversity. Demanding social equality between men and women and between the different "races," albeit fruitful and important, also raised questions. Equal to what or to whom? Who or what is the universal standard of modernity and emancipation? In addition, what about the differences between people's lives and the appreciation of those differences? Ever since the late 1980s, pluralism, multiculturalism and diversity have become the answer to bypass universal claims.[36] Here I want to argue that neither egalitarianism nor diversity provide in themselves stable ground for feminist and anti-racist politics, and I will consider what can be learned from my analyses of genetics, especially for a politics of diversity. Let us, therefore, turn to the narrative on *diversity*.

Within gender studies, it has been shown that the gender or man–woman category is discursive and heterogeneous

Ever since the nature–culture debates in the 1970s and the appreciation of the sex–gender distinction as such in the 1980s, little attention has been paid to "biological" sex. Differences between men and women were best understood as cultural, and gender became the field of studies, debates and interventions. Some scholars have, however, suggested that the issue of "biological" sex is much too important to be left in the realm of scientific discourse only, and much too complex to be treated as a stable reference.[37] They have put forward the argument that there is no reason to apply a special treatment of the biological as being different from the cultural.[38] Equally one cannot presume a given sex–gender distinction: even if the very distinction performs itself.[39] Rather than what is essentially there in biology, the angle they suggest is an examination of practices in which sex materializes, is enacted and made to matter.

In this study, I started out with the tantalizing account from geneticists that the boundaries between populations may shift depending on whether mtDNA or nuclear DNA is used to investigate these. Population is not stable. Soon after this, I learnt also that the difference between the Y-chromosome (nuclear DNA) and the mtDNA was perceived as a sexualized system. So in studies of lineage, it is not the distinction between XX–XY that makes the sex difference, but rather that between mtDNA and Y-chromosomal DNA. An interesting matter in its own right, since both males and females carry mtDNA. So how are sex differences "done" in genetics? And where can these be located? To answer these questions, I have looked specifically at research where Y-chromosomal and mitochondrial studies of populations were compared.

The leading questions in my analyses were how are sex differences performed in such studies and where can they be located? To answer these, a distinction was made between three different practices in studies of genetic lineage: the practice of genealogy, the practice of DNA and the practice of genetic lineage. These could cautiously be seen as the practice of theory, the practice of experiment and the practice of analysis. Where sex difference is located and how the sexes were performed differed in a significant way between these practices. In the practice of genealogy, the sex of individuals is irrelevant to genetics, and genetic sex is performed as a pattern of inheritance: whether a specific DNA fragment is passed on to the individual by the mother or the father. In a practice of DNA, genetic sex may be irrelevant altogether, such as in mtDNA, because both males and females carry this type of DNA. However, it may also be actively performed, as in Y-chromosomal DNA research. There the difference between males and females becomes important because females do not carry a Y-chromosome. However, in such a practice, geneticists were not interested in the individual as such but in collections of samples or of populations. They, therefore, studied the samples which were to hand and which worked, given the availability of techniques and time. Hence, in the handling of DNA, genetic sex was not so much performed as a quality of a *sampled individual* but rather as that of an *individual sample*. In addition, in a DNA practice, sex difference could be enacted in various other ways. It may be performed as the spatial division between samples, as the information contained in records about the samples or as repertoires from other practices that had entered the laboratory and had become part of the routine. Consequently, both during experiments and in theories about DNA inheritance, genetic sex was hardly ever performed as a quality of an individual. Rather sex difference materialized in various technologies of studying similarities and differences. Does this then mean that genetic sex is never enacted as an individual quality? In my examination of the practice of genetic lineage, the practice in which the data are analyzed and put in the context of population history, I have argued that the various different versions of the sexes that could be found in the laboratory were subsumed. In the cases of mtDNA and Y-chromosomal DNA research, it was shown that the data, put in the context of population history, could stand for women and men. One could say that, in addition to the standardization of DNA as a technology, the naturalization of sex difference also enables the naturalization of population and differences between populations.

So what can be learned from this investigation? A focus on routine practices and on what scientists *do* showed how sex difference materialized in technologies. It also showed that different versions of the sexes might circulate in laboratories. Next to sex difference in terms of man and woman, geneticists had many other technologies to hand in which it could be performed. Moreover, the sexes might be performed and made relevant, but such was not always the case. Performing them, no matter how it was done, was temporary to the point of being unstable.[40] Hence, in laboratories, genetic sex is not so much an essential quality of an individual but rather the effect of technologies and practices. However, to state that technologies and practices are heterogeneous and that various different versions of the sexes can be found in laboratories is not to suggest that sex is a list of endless references to something essentially there in the DNA. Rather, my analyses show that genetic sex itself is a matter of technology and that specific versions of the sexes can be performed in certain

practices but not in others. This indicates that scientific practice itself questions
the very distinction between sex and gender. If sex is not one thing but rather
manifold, and if sex is a matter of cultural practice, it invites us to rethink the
natural and the cultural.

In debates about race and racism, the distinction between nature and culture had a different aspect from that in the debates on sex and gender

As a way of coming to terms with racism after the Second World War, race
was put on the agenda of a UNESCO meeting in December 1949, resulting
in two statements. The first, presented mainly by sociologists, psychologists
and cultural anthropologists, suggested that there was no scientific basis for
human biological races and that race was being mistaken for population.[41] The
second, which was a consequence of geneticists' and physical anthropologists'
discontent with the first, especially on the issue of presumed innate intelligence,
turned the argument of scientific evidence around. The very lack of evidence
was used to propose further research on race.[42] A group of 96 scientists were
consulted before the second statement was released, and a number of them
suggested ongoing studies and debates rather than a final consensus.[43] Nevertheless, as a spin-off to these debates, population became the preferred category
of biological research and race was confined to the realm of ideology and bad
science.

As has been pointed out, the discourse of the Diversity Project is centered
around "population." The history of race since the Second World War suggests
that there are stakes in studying practices of population. Race is but one category
of making differences, and naturalized versions of population might do the
work of essentializing these in a rather similar way.[44] What impact, however,
do current technologies have on how race or population is enacted and how is it
done? Specifically, since population comes in various versions, its naturalization
suggests that something is also happening to race. The ever-growing number of
genetic markers applied in the field of population genetics increases the ways in
which population can be enacted. As had been indicated, the boundary between
one population and the other may be very different, depending on whether
populations are studied on the basis of nuclear DNA or mtDNA, or whether
the clustering is based on a small or a large number of genetic markers. This
indicates that neither population nor race is pre-given and suggests that race does
not necessarily map on pre-existing biological classifications.[45] This practice
of race is not based on a pre-fixed category into which individuals are fitted.
Rather it works the other way round. The starting point is an individual with an
endless amount of genetic information through which race can be enacted again

and again as something different.[46] One could say that, for better or worse, this aspect of new genetics is producing populations, races and sexes in excess. The question is, which versions do we want to live in?

Confining race to the realm of ideology and bad science has produced problems, and not only for scientists interested in biological classification

The treacherous, hotly debated and slippery field of biological race also seems to have become a "no-go" area for scholars of science and technology.[47] This is indicated by the virtual absence of research on the materiality of race in techno-scientific practice.[48] There might be various reasons for this, from simple lack of interest to problems of addressing the materiality of race without fixing it.[49] Could it be that in addition to such reasons, the very absence of a nature–culture divide vis à vis race, like the one attributed to sex and gender, has contributed? Whereas the field of gender and technology has become established in STS, studies of race and technology are virtually absent. It seems that the impossibility of referring to race, just like sex, without referring to something fixed in "biology" has contributed to the omission. Gender appeared to be productive for those who wanted to avoid the materiality of biology; race, however, has lacked such a cultural counterpart to do a similar job.[50] Race remained captive in the realm of scientific ideology. Rather than an issue of investigation it was treated as an issue of debate. In many cases it was debated for very good reasons.[51] However, as many feminist scholars have argued, the very distinction that confined sex to biology has led to problems in thinking of the body and its materiality in technologies. Hence, there is an urge to start to take race into account in studies of science and technologies and how it materializes in such practices. In addition, there is an urge to remove race from the domain of taboo and deviancies in order to investigate how it is practiced on a routine basis. For race does not only matter in terms of hierarchical distinctions, not only in terms of inferiority and superiority,[52] but in many other ways. Among many others, race might "matter" as a difference between the isolated population and the genetic melting pot, or as a difference between what is genetically proximate or distant, or again in terms of geographical clusters of variability. In my studies, I have focused on technologies of making such distinctions and traced how race is embodied in them. Taking routine practices into account, it was shown that similarities and differences are neither vested messages in the DNA nor ideological additives of scientists. Rather, racial distinctions materialized in technologies and practices.

Hence neither race nor sex can ever be simply biology, or simply ideology, just as they can never be simply nature or simply culture. They refer by "nature" to the socio-material density of that what we call biology. Here lies the point in stating that the "nice thing about DNA is that everybody has it."[53] What anybody may "have" is indeed a matter of doing. Now that biology has taken the shape of DNA, genes and genomes, testifying to the contaminated objects called nature–culture, race and sex forces us to take account of how biology is done. However, the stakes for feminist and anti-racist politics lie in denaturalizing both DNA and technology simultaneously.

On the multiplicity of population, sex and race and the interdependence of nature and technology

My analyses have centered around technologies, which enabled me to account for the fluidity of practices and the performed quality of objects. At various places, I have suggested that the multiplicity in nature should not be taken for a list of references to something essentially there. This also has consequences for a politics of diversity.

Diversity has taken on a normative aspect since the early 1990s: it carries a positive meaning, so it seems.[54] It does that in the very sense that it mobilizes a critique of homogeneity or easy classifications (such as self and other, working class and upper class, man and woman, and the like). However, it does not self-evidently prevent the foundational power of more refined categorizations. Indeed, it does not prevent such powers without the mediated and thus temporary nature of categories being in focus. Here lies the importance of what Donna Haraway called embodied vision.[55] Her metaphor of vision does not imply that seeing is directed by minds, nor does it suggest that some bodies are better equipped and generate better sight. Rather, the focus is on the dependent nature of both seeing and that which is seen. It aims at technologies in practices. Hence, both viewer and viewed are effects of technologies located in time and space. From this we learn that there is no one stable ground, not even many, for opting for a diversity or egalitarian politics. This is true even if there might still be very good reasons, for specific people, in specific places and at specific moments in time, to choose either of the two. Thus introducing diversity into feminist and anti-racist politics does not, as such, yield better politics. Introducing more axes along which social, technical or natural phenomena are studied does not do the job. For peoples' lives cannot be understood in terms of an addition of coordinates. Diversity as such does not challenge the idea that there is an allegedly stable reference for identifying similarities and differences. From practices of genetics, we can learn that the ground for politics is crafted by technologies of similarities and differences. The core issue in such politics is how are people performed as similar or different? Which technologies have pride of place in producing diversity? And to what effect? Meanwhile, politics involves knowing that similarities and differences are neither the beginning nor the end, but rather the fluid space of technology, blood and DNA in between.

Talking forwards to politics

To conclude, let me return briefly to the Diversity Project and its mapping capacity. Making maps is making links. The spatiality of geography produced in maps transforms relations and determines what is near and what is far away on a map.[56] As Marilyn Strathern has put it, maps have "both a domaining and a magnifying effect."[57] They are political objects, not only because of the boundary work they do but also because they produce visual centers and margins. A genetic map of the world embodies these politics as well. In the discourse of the Diversity Project, DNA is placed in the realm of nature. Among many other things, this discourse is about conserved genes and isolated populations versus mangled genes and the Western melting pot. Those who are not considered to be connected to the global traffic of humans and things, especially those in faraway places, carry DNA that is considered a source for understanding how the melting pot must have come about. This is the kind of mapping that is familiar to us by the virtues and vices of history.

In this concluding chapter, I have spent time mapping out the practices investigated in my study, and the different insights gained about genetic diversity. I have traced three narratives: three ways of talking about how to do politics with DNA. Talking about naturalization, I have argued that in genetic practice there is no such thing as a natural object: rather, the objects of genetics are enabled by technologies. However, the naturalization of technologies has the naturalization of genetic objects as its effect. My point with naturalization is that both objects and technologies appear autonomous and detached from the practices in which they are produced. Indeed naturalization reifies the distance between the "isolates" out there and the "technology" here. Hence to do politics with DNA is to take into account the practices in which it is studied. In the narrative about standardization, the focus shifted from the *how* to the *content* of practices. My aim with this is to address the normative content of standards. For this content tends to be obscured, thereby obscuring the normativity of objects enabled by such standards. Geneticists show special interest in difference. However, the process of standardization makes inaccessible the practices (built into technologies) that are applied for making differences. Genetic differences thus seem neutral and facts of nature. Obviously, the politics of such differences is the real matter of concern in the Diversity Project. In my narrative of diversity,

I focused on sex and race and argued that similarities and differences are not inscribed in the DNA. Rather these are enabled by the technologies that help to produce them. Addressing sexual and racial differences in genetics, my aim was to point to how these become part of routine technologies. Given its crucial role in the Diversity Project, I have specifically put forward the argument that race should be taken out of the realm of ideology and deviancy (i.e. bad science) and that further investigation is needed to learn more about how it is practiced in scientific routines.[58]

Finally, this book can be seen as an attempt to *denature* the objects of genetic diversity, whether these are individuals, population, race or sex. It is also an attempt to show how different versions of these objects are coordinated and performed as one: often as the sticky and crusted versions that we have become familiar with through history. This implies, for instance, that population should not make us feel more comfortable in the sense that as long as population geneticists are doing population and not race, than the world is fine. Race is, as Margaret Lock has argued, "one special case way of lineage making."[59] Population is, among others, a second case. Both race and population are conflated with culture and biology and both can be enacted as essential qualities of groups of people. There are stakes in observing how that is done and for what purposes. There are stakes in taking account of the routines of making similarities and differences and the normativities they involve. Finally, there are stakes in investigating how the differences of genetics matter (materialize) in the world outside. Such an awareness may also sensitize STS scholars, geneticists and gender and anti-racist scholars to the types of links and lineages which go to constitute the map of the Diversity Project.

From margin to center

This book spans a period from approximately 1991 to 1999. What has happened to the Diversity Project since then? It has developed along different paths. One of the most tangible achievements of the project is the production of a databank in 2002 consisting of 1065 human cell lines from 51 different populations.[60] This databank is a compilation of cell lines that had already been developed in different cell line consortia of the Diversity Project. Since 2002, they have been moved to the Foundation Jean Dausset (CEPH) in Paris and fall under the supervision of Howard Cann. To avoid (commercial) exploitation of the cell lines, only DNA samples will be distributed to researchers: "DNA from the LCLs [lymphoblastoid cell lines] will be distributed to investigators who agree to type all the DNAs with their genetic markers and contribute the results to a central database. LCLs will not be distributed."[61]

One could say that this databank is now the only internationally coordinated outcome of the Diversity Project. Other than this, the project seems to have failed to get started as an international activity.[62] However, whereas it did not manage to get sufficient funding for an international initiative, many associated laboratories (in the USA as well as Europe) are receiving financial support for their studies of diversity and population history, and many of these researchers are collaborating on various studies. Moreover, the use of databanks (such as the mtDNA databank and the Y-chromosome databank) to which new data are continually supplied from different scientists, have made international collaboration lighter work. Thus, instead of one project, the Diversity Project had become "a collection of networked efforts."[63]

Given its controversial history and compared with the HGP, the Diversity Project had over the years occupied a place in the margin in terms of funding, support from academic communities and from political actors. By contrast, *diversity research* had moved from the margin to the center for several reasons: research goals of the Diversity Project had been exported to countries outside of Europe and the USA, such as India and China;[64] different laboratories working on genetic diversity and the history of population managed to receive more funding for their work; knowledge developed in such studies, for example on Y-chromosomal variation or on mtDNA analyses, can be widely applied in forensic DNA practice. Most importantly, diversity research became pivotal because studies of diversity have now been embraced as a valuable means to further understanding of "the human genome" and to produce genetic knowledge that could be applied in the comprehension of diseases. The US National Institute of Health (NIH), has launched two well-funded and related projects in this respect, The Genetic Variation Program and the Environmental Genome Project. The first, funded since October 2000, will study common diseases such as high blood pressure in relation to genetic variation and will develop tools to analyze polymorphisms in the respective genes. The second, established in 1998 and funded by the NIH National Institute of Environmental Health, is a multiyear and initially a 60 million dollars project aimed at studying about 200 genes in US populations to investigate the genetic variation (polymorphisms) of these genes in relation to disease susceptibility and to environmental exposure. A second goal of this project is the compilation of a single nucleotide polymorphisms databank.[65] Although the issues of population and race are understated in these projects, they are at their heart. Thus while moving from margin to center, genetic diversity research had become empowered with a discourse on health and illness, and supported by money and politics. It has consequently become even more urgent to pay close attention to its practices of knowledge production. At stake in genetic diversity is not only lineage or descent but also normalcy or deviancy: good genes or bad genes.

On method

Patchwork methods

It is a well-kept secret that the methodological section of academic writings is produced with hindsight. Yet as students we were all taught how to write academic papers: first choose a topic (preferably a topic that makes your heart beat faster, or maybe one that does not); then develop a hypothesis and design some research questions; finally choose and describe the proper methodology to tackle these questions and to prove the thesis right or wrong. After this you can start to do the work of research. Traditionally the methods form the first chapter of the paper. Along the way, your advisor may tell you not to spend too much time on this part, because your research material might surprise you and you might have to change your methodology.

There might be didactic reasons for the appearance of the methodological section at the beginning of an academic text. It is thought helpful to know from the start what the writer is on about and how she or he will tackle the problem dealt with in that work. The methodology section thus functions as a road map of what is to come. This might be true in some cases. However how a text is read can vary. There are different ways to situate it. Some flip through a text, to choose one particular chapter to read first; others like to glance over the footnotes and references, and yet others look at the index to develop a first idea about the topic. In fact, different texts might do different things for a reader and might invite different handlings. Think of the difference between a text that contains pictures, diagrams or ethnographic material, and texts that do not.[66] So in many cases "method first" may be more a matter of convention than one of didacticism.

Does this then mean that method is irrelevant? No, but there are serious problems with "method first." It reproduces a standing notion of method as a toolbox. One that is made suitable for the field about to be entered. There, the object of study seems to be waiting for the researcher and his/her toolbox to reveal it. This notion presents method as detached and instrumental. Method does not intervene really with what is to be studied. It simply makes visible what would have remained concealed. In fact this notion of method has been criticized quite extensively in feminist and anti-racist studies of science. Method is not neutral, it was argued. Different repertoires about the social order outside the sciences may enter the laboratories and function as explanatory schemes of the natural order. Method, so it has been argued, reproduces certain normative distinctions, such as the passive egg cell and the active sperm (mirroring alleged ideas about women and men in society).[67] This critical approach also suggested that different methods might provide different insights about a particular object. From this perspective, an object does not have one side but many. Any one method would place particular aspects of an object central stage. This is specifically important because of the privileged status of scientific knowledge: the idea that what is revealed by scientific inquiry is granted a higher status and produces a hierarchy between itself and other aspects of an object. To give an example. It has taken scientists quite some time to learn that the egg cell is not the passive receiver they held it to be. Even though it does not have to swim its way up to the sperm, it is active. For it is very much involved in which sperm will enter for fertilization and under which condition. Furthermore, it seems that it is not the fastest sperm who will become the supreme candidate, but probably those who seem less active because they move slower.[68] So methods are not neutral and they produce privileged accounts of the world at the cost of many others. However, there is more to the "method first" problem.

What if method does not just produce a particular account of an object but in fact produces the object itself? What if method is actively involved in enacting the object? And what if such an object does not pre-exist the intervention? Here is, for example, what Bruno Latour and Steve Woolgar said in this respect: "Without a bioassay, for example, a substance could not [be] said to exist, the bioassay is not merely a means of obtaining some independently given entity; the bioassay constitutes the construction of the substance".[69] When ethnographers entered the laboratories in the 1970s, their major aim was to locate scientific knowledge somewhere.[70] Contrary to the idea that scientific facts are universal, detached from the scientific activity itself, and products of a rational process, they wanted to give flesh to these facts. Their concerns were in line with Thomas Kuhn's observations on, and questioning of, the cumulative and rational nature of science.[71] They thus turned their attention to how science is done in everyday practice, and how scientific facts get made in such routines. Obviously the reports of scientists on their work, such as in the "material and method" section of scientific papers, did not give many hints about that. One could say the accounts on methods therein were submitted to the first problem of what I have called "method first". Now, without rehearsing what different laboratory studies have done, it is important to note here what had happened to method. First of all method became methods, in the plural. In laboratories, many different techniques and procedures of making science were at hand. Second, natural phenomena – objects – are not simply placed on the table for inspection, but they are enhanced and upgraded as to make them fit the fabric of the laboratory, and aligned to technologies to make them studiable. This entanglement of objects and technologies continues along the process of inquiry. Objects are dependent on and enacted with the help of different technologies (including the technology of writing). Again Latour and Woolgar: "It is not simply that phenomena *depend on* certain material instrumentation; rather, the phenomena *are thoroughly constituted by* the material setting of the laboratory".[72] Therefore, to study, with whichever methods, is to intervene. Methods act, and they help enact the object. Moreover, this process is not straightforward or firmly outlined from the start.[73] It usually encompasses tinkering and changing methods along the way. The outcome of scientific work is never certain. What can be claimed for the methods of science can also be claimed for the methods of those who study it. Obviously it holds for this book as well. The methods and objects in it emerged in process. They have shifted and changed along the way.

Being there/here

There are several reasons why I ended up studying the Diversity Project in laboratories. Quite soon after I had heard about the Project, this was in 1995, I interviewed a Dutch geneticist who through his work for HUGO was familiar with the initiative and its aims.[74] Although I had already gathered some information about the Diversity Project and genetics in general, this interview, as initial interviews often do, generated a lot of material. One issue mentioned in it was important for the direction of this research. Through this conversation, I learned that the boundary between one population and the other is not stable. Different parts of the DNA may produce different accounts of what population is: nuclear as opposed to mtDNA was the instance given. This stimulated me to learn more about how population is "done" and what kinds of knowledge and technologies are implicated in that. I was familiar with some of the laboratory studies that were published. My guess was, therefore, that it should be possible to study how objects get made in scientific practice and what kind of socio-materiality they embody. So this was the first reason why I decided upon a participant observation in laboratories.

The second reason was because of a kind of morality that says that "you have no right to speak unless you know what you are talking about." And I did not know much about genetics. Ironically enough, I learned in the laboratory that there are many ways of knowing and thus many rights and reasons to speak. For example, after I had given a talk in one of the laboratories about population in genetics after the Second World War, different books and documents, such as Daniel Kevles *In the Name of Eugenics* or the UNESCO document on race, circulated in the lab and were discussed during lunch or coffee breaks. Moreover, it turned out that the position of a novice is a good methodological tool.[75] Since I had to learn to do the experiments from the start, it allowed me to unpack the techniques and routines. Any experiment that did not work brought to the surface the different components involved and their respective role in the process. This position also forced my colleague laboratory members to be explicit about these and to "think out loud" when we tried to solve the problems that occurred.

The third and major reason had to do with the Diversity Project itself. Although I was both alarmed and troubled by the initiative, I was wary of subsuming the project in a general critique, for example one of "imperialism" and racism in science. Besides, why would this project be "bad science" whereas others were not? For as Donna Haraway has stated: "It has proved easier to slow down or stop the HGDP [the Diversity Project], a kind of oppositional effort, than to question the powerful HGP itself. That makes the trouble with "difference" built into this potentially positive scientific project all the more disturbing – and important."[76] So the Diversity Project is political to the point of being controversial. However, I wanted to move away from the notion that takes politics merely as a dissonance between social groups or as incommensurability between different perspectives on an issue.[77] Rather, I wanted to trace it in the materiality of everyday routines. So instead of focusing on the controversy and the debates that surrounded it, I chose to study locales where nothing strange seemed to be going on: locales of science as usual. However, in the laboratories, I hit upon unexpected implications of the controversial status of the Diversity Project.[78] It appeared to me that geneticists I talked to did not want to associate their work with this project. Even though they were studying genetic diversity and even though their names would appear on the "International Executive Committee" list or on other documents, they would speak of the Diversity Project as something disentangled from and external to their daily work. So I wondered, "where is it?" I was, of course, aware that the Project did not get started officially, that is, that it did not receive the funding that its organizers had asked for. But it was initiated and both HUGO in Europe and the US Science Foundation supplied some funding for a series of workshops to assess its potentials. It was in the process of studying it in the laboratories that I realized that these workshops were already doing the work of organizing the scientific community. In these laboratories, the Diversity Project was both absent and present. Because it was controversial, geneticists themselves did not articulate their ties with the Diversity Project. However, these ties were articulated in the organization of their work. Through the workshops, scientists started to work together, to exchange technology, samples and people and, in fact, to enact the

Diversity Project. This implies that the field of study for this book was not only a locale where I did my research (the laboratories and conferences I visited, the interviews I held or documents I read), but it was also a *political* site. This site, in which the organization of scientific work was taken into account, made actual a project that, one could say, did not exist. It performed the Diversity Project as something that *is*, rather than something to be, sometimes in the future.

Lab F in Leiden offered me a training in some of the basic tasks of a technician. When I first visited this laboratory, it was a late afternoon and the laboratory members were having a tea break; we had a conversation about their work. While we were talking about what they do there and how they do it, the head of the laboratory kept referring to their work as "simple" and "routine." So I dared to ask the question: "Could I, for example, also learn how to do it?" "Of course," he answered, "do come by and we'll teach you." After a couple of months, I returned to this laboratory. Over a period of three and a half months I combined the training with a study of the laboratory itself. Together with the head of Lab F I attended a conference on the Diversity Project where I met many of the scientists participating in the project.[79] At this conference, I also met the head of Lab P Munich.[80] In 1997, I spent six months in this second laboratory and participated in one of the projects in the field of population genetics. These laboratories differed considerably. The first laboratory's core business was the production of forensic DNA profiles; in the spare time, research projects were developed. It was an ordered space where five people were working.[81] The second laboratory was an international research laboratory with many visiting researchers, all interested in one way or another in genetics and evolution. This had implications for a novice participating in the research. Whereas the first laboratory was an ideal place for a beginner; in the second I found myself on my own. Not that there was no supervision or colleagues to help whenever that was necessary; on the contrary. Rather it was the technology on which it was difficult to rely. Since many different researchers, also from neighboring laboratories, used the same machines and chemicals, hardly anything was standardized. If one had designed a program, say for the PCR machine, it may no longer be there the next day. This forced me to take much more account of the technologies and chemicals and how these were designed.

During my participation, I wrote down my observations either in the laboratory itself or in the evenings at home, and conducted interviews with members of both laboratories at the end of each study. In gathering published papers, I was struck by the generosity and involvement of laboratory members in bringing some of them to my attention and for keeping me up to date, even after I left the laboratories. Having been engaged in laboratory work made it easy to become "a member." However, it also imposed some constraints upon my fieldwork.

First of all, the temptation is to go epistemically native. A major reason for this is that a laboratory environment imposes a specific type of normalization upon those who work there. The very fabric of the lab demands a kind of subjectivity centered around the pace of the work, the planning of experiments, the talks, which are often about problem solving such as machines that are overbooked or not working, or about how to get the data and when to write down the results. Once I became familiar with the various projects, it proved difficult to relate to them other than within the conditions of these practical concerns. In addition, several times during my research, participation stood in the way of observation. Often there was simply no time to take proper notes or to be where the action was.[82] My main focus at such times was to get the results, to make things work or to establish the conditions for the experiments.

Yet as an observer one is also tempted to neglect these practicalities of research and to develop a kind of science critique instead. In a sense, it is tempting to tell the strange stories back home without bringing along "the laboratory." However, my experience was that laboratory members were themselves aware of the social aspects of genetics, especially of racial issues. As Susan Leigh Star has observed: "As a group of respondents, scientists are particularly difficult and rewarding because they have often thought rigorously about the issues we are investigating, and about which we are ourselves uneasy."[83] The geneticists I encountered were self-reflective upon their work and the particular environment in which they carry it out, and they were capable of standing apart in order to develop a more sceptical view. In a significant sense, this helped me to center my analyses around technologies and everyday practices and not to impose pre-set categories on the kind of work they do. Another and related point is that in some ways one can never really leave the laboratory. My experience is that the position of both participant and observer remained intact. This became apparent during the many visits I paid to the laboratories after I had finished my field work, in the various personal contacts that I maintain with some of the laboratory members and in the material objects that I brought home, such as my (observer) field notes and my (participant) journals. Hence participation and observation continued in parallel during the process of writing and had to be negotiated in various drafts of the chapters. While the ties which I developed with the laboratories may be particular to my studies, the point itself is, however, more general and methodological. I will, therefore, expand on it.

There is a certain epistemic quality to the phenomenon of participant observation. It disturbs research design, time schedules and methods set out for gathering the material; something probably common to all research. However, it does more. Participant observation requires the researcher to go out to study the "other culture," yet it disturbs the very distinction between the field, there, and the writing, here.[84] This blurring of boundaries in the end-products of participant observation (i.e. in written texts) has been brought to our attention by ethnographers such as Clifford Geertz.[85] The anthropological "field" as a site on which difference had been mapped has made room for a semiotic one of which the objects and categories have moved well beyond it.[86] However, I wish to point to another aspect of participant observation and elaborate on the epistemic quality mentioned above. After I had finished my fieldwork and went home to do the writing I was confronted with the field once more. It was right there, on my desk. Not only had there been DNA samples in my refrigerator, "gel Polaroids" in files, but also the field notes, documents and papers appeared to be much more than artefacts from another world. Once some of the material had found its way into one of my chapters, it started to do its own work. At some points it refused theorizing and it refused even to get out of my texts again. Now and then it urged me to go back out *there* and learn more about it, in the library, in MedLine or in the laboratory. It occurred to me that the complexity of the locales I had left behind had traveled all along, not only with me because I had been there, but particularly with the material. This means that the material seems to "*extend*" not only the field but also the researcher.[87] The researcher is performed both as participant observer and as writer. To put it differently, in the process of writing the researcher is "*interpellated*" by the ethnographic material and placed *in-between* these positions.[88] Thus, it is not only in the final texts of ethnographers that the boundary between the field and the writing is blurred, but in a very material sense also *during* the writing because of the very capacity of the field to move to other places in the world via such ethnographic material. It is in this sense that my statement that one can never really leave the laboratory or field should be understood. It might also be for this reason that ethnographers have grown to be squeamish about their material since it performs them as such, and it always bears with itself a world that wants to speak, often with many voices.

Does this mean that the material has presented itself or the world it came from? Does this mean that writing is without theory?[89] Even though "the field" was on my desk, it was not there by itself. There were also theories in the form of texts: books and articles from the field of STS, gender and anti-racist studies, but also philosophy, anthropology, cultural studies and genetics. They dealt with bodies, gender, technologies, gifts, cultures, race, hormones, double helixes, genomes and blood – among other things. Both material and theory had to be negotiated in the process of writing. The final text of this book is an analysis and not a description of what the field is like or how it can be found out there. The narrative of the previous chapters evoked a distinction between ethnographic accounts and their analysis. This might be read as a distinction between the reality of the field and reality of writing: the analyzing and theorizing of the material. However, even though the ethnographic accounts are faithful to the material I gathered, these too are assembled, framed and guided by theory. They are thoroughly theorized. As I have stated, the material had to negotiate its place in the final text. In addition, even if the references appeared mainly in the endnotes, the theories did their work in the body of the text. They were part and parcel of my analyses.[90]

Partial lineages

As Latour has taught us, the fate of any text is in the hand of its future readers. Where, however, would a potential reader visiting a library find this book? On which shelf would a librarian eventually place it? Let me speculate on that a little by drawing some partial lineages between this text and other ones. These lineages are aberrant yet real and, therefore, partial. It is a kind of lineage that is not dependent on one origin. Just like the genetic lineages analyzed here, they are fabricated and enabled by various different techniques and points of departure.

Because of a focus on the routines of science, some lineages can be established between this text and the laboratory studies. Just like these studies, its main object is what scientists *do* when they produce knowledge. In this doing, scientists are not by themselves. They are in contact with colleagues in other laboratories, but also, closer to where they work, they are surrounded by technologies, texts, theories and the material they study. Ethnographers have taken these materialities as well as the rites and rituals of scientific work seriously while analyzing what counts as, or what is made to be, good scientific knowledge. Science, as these studies have shown, is a cultural practice.[91] Its product is a material culture where objects, technologies and humans work together and give particular meanings to such an environment. Despite its contextual and situated nature, and even though scientific work is not about doing the same thing everywhere, it involves the stabilization of methods and processes of that work. Various studies and concepts have been brought forward to understand processes of stabilization and standardization: for example in terms of immutable mobiles, oncogene bandwagons, boundary objects, stabilized networks. In these studies, standards are not viewed as *matters of facts* that arise naturally, but rather as local achievements. They have to be established again and again while moving across practices. So given all this, a librarian might chose to place this book on the shelf of laboratory and associated studies in the field of STS.

However, there are other links that can be made and lineages that can be established. What if the topic of this book would be extended? Not merely scientific practices and routines but also the object of science: sexes, populations and races, among others. Placing these objects central would mean shifting fields, or shelves. We move from the shelf of the laboratory studies to that of gender and anti-racist studies of science.[92] There we find studies that have examined and traced biases in the language or discourse of science, giving insight into hierarchies in the designation of agency to naturalized categories.[93] This may be a hierarchy between the races, the sexes or between racialized or sexualized entities that do not necessarily coincide with human individuals, such as the wild type versus the mutant, the active sperm versus the passive egg cell.[94] Moving along this shelf, we would also find studies that investigate the social groups that do scientific work, showing a male bias and revealing the contribution of women and, occasionally, that of men and women of color.[95] Yet others have the scientific method as their topic. They consider the Eurocentricity and masculinity of scientific methods. These studies demonstrate that these methods frequently involve a distinction and a hierarchy between a (masculine) *subject* of research, namely the scientist, and a (feminine) *object* of research, namely nature. Hierarchies also exist between culture, as an achievement of Western science, and nature, as the naturalized and pre-given non-West.[96] The books on the shelf of gender and anti-racist studies of science show a particular interest in the effect of knowledge for the object of research, and not infrequently this object is the female or colored body. This book is sensitized by these and related studies to the power and politics of science. In line with these studies, it has placed the object of science center stage: to shed light not merely on knowledge practices (such as in the classical laboratory studies) but also on the objects. Such objects would travel in and out of the laboratory, changed, enhanced and upgraded along this process. For as Donna Haraway has argued: "The form in science is the artefactual–social rhetoric of crafting the world into effective objects. This is *a practice of world-changing* persuasions that take the shape of amazing new objects – like microbes, quarks, and genes."[97] Thus the objects of science matter not only within but also outside of the laboratories. The shelf of gender and anti-racist studies of science would, therefore, be a second possibility for placing this book.

There may be more. I will mention just one further possibility for establishing partial lineages. This can be done by taking the politics of scientific practice a little further and is yet another way of extending the topic of this book, from what Annemarie Mol has called a politics of *who* to a politics of *what*.[98] First of all, there has been a tendency to treat the politics of science as deviancies to a normal process, specifically when the issues are race and sex differences.[99]

By contrast, this book, and it is by no means the only one, examines politics as an integral part of "science as usual" and views how such politics get built into standardized technologies and laboratory routines. It does not focus on what geneticists think, nor on how they talk about their objects, such as sex difference or race. The aim is not to unmask geneticists as being racist, sexist or biased in any other sense. For the point is not so much *who* is conducting science as *how* is it conducted. So the politics and normativity of science is to stay. It cannot be removed to acquire a neutral field. Secondly, focusing on how science is done emphasizes that it produces different versions of objects. These versions are normative, and enacting one at the cost of others is not necessarily bad, but it is a political matter.

So to which shelf are we moving then, if we are interested in the normativity and politics of scientific practice? Well, there is a relatively new branch of STS in which normative issues combined with an interest in scientific practice are being addressed.[100] It is a branch of STS that borrows from cultural studies as well as the anthropological studies of science. It may even be said that it consists partially of these studies: but let us not be purist about that. In any case, on this shelf we find studies that pick up and readdress classical normative questions, such as: how does science and technology change social worlds and for whose benefit? How do social worlds get built into technologies? What kind of politics do technical objects carry with them? How do they affect the ordering of the world and processes of inclusion and exclusion?[101] In particular, the books that deal with medical practices, genetics and the new reproductive technologies pay attention both to how scientific facts are assembled, made and consolidated and to the morality borne by both objects and technologies.[102] They raise questions concerning normalization, naturalization and standardization, and they show how personhood, gender or the body are locally achieved.[103] Even though most of these books do not have the laboratory as their site of study, and others may have it as merely one of the different sites where knowledge is made (such as the clinic, the consultation room, the doctor's room, etc.), this book would be in good company on this shelf. It is, however, very well possible that the librarian would despair and reason, since it is dealing with genetic, why not place it on the ever-expanding shelf on the social aspects of genetics. Again, this is just one possibility among many.

Notes

1. On the notion of contrast as a methodological tool, see Mol (1991b).
2. Questioning the ontological difference between micro and macro, or local and global, phenomena or mechanisms is, in fact one of the major achievements of the "actor network theory," see for example Law and Hassard (1999). For an elegant

elaboration on the relation between the local and the global, and how the local and the global together specify the limits of a field, see Strathern (1995a: 167).

3. Although not very satisfied about it himself, John Law (1994: 188) has termed this type of work "a kind of juggling while trying to keep all the balls in the air."

4. See for some other examples, Pasveer (1992); Mol (2000, 2002); Law (2002).

5. On coordination, see Callon (1991); Galison (1999). In her ethnographic study, Annemarie Mol (2002: 53–85) traced two forms of coordination through which different versions of atheroscleroses are drawn together: the adding up of test outcomes and the calibration of test outcomes. John Law (2002: 15–32), studying the military air-plane TSR2, discerned a long list of strategies of coordination, from the physical structure of a brochure about TSR2 to the navigation system described in that same brochure.

6. Translation is central notion in the actor network approach to science and technology and akin to the analyses conducted in this study. See for some examples Callon (1986a,b); Latour (1987); Law and Callon (1997). See also Strathern (1991: 53).

7. See Callon (1986b); Law and Callon (1997). On script, see Akrich (1997).

8. The same can be claimed for various other objects or technologies, see for an example Rheinberger (1997).

9. See also NRC (1997; especially pp. 5 and 39–41). In this document, which assesses the potentials of the Diversity Project, it is stated that geneticists should not aim at compiling a priority list for markers. Which markers are good depends on the goal of research. However, and somewhat contrary to this, the evaluating committee expected bi-allelic markers to become the markers of choice in the near future.

10. For a similar observation about research devices, see Rheinberger (1999: 420).

11. Fujimura (1987).

12. This is a slight modification of Latour's first methodological principle (1987).

13. For some examples, see Bosk (1992); Kevles and Hood (1992); Ginsburg and Rapp (1995); Cunningham-Burly and Kerr (1999); Novas and Rose (2000); Rapp (2000).

14. Obviously this is an important point of many STS studies. It might seem to contradict the statement of Bruno Latour (1987) about scientific facts and their future users. However he is making a different argument, namely that scientific facts can never be end-products.

15. Such as *The International Planning Workshop*, held in Porto Conto, Sardinia in September 1993, and *Human Genome Variation in Europe: DNA Markers*, which took place in Barcelona, Spain in November 1995.

16. See for example, Star (1991).

17. See also Reardon (2001), who argued that the problems that are met by the Diversity Project have to do with the failure to "co-produce" the natural and the social order. This is specifically apparent with reference to the issue of population.

18. See, for example, Clarke and Fujimura (1992a).

19. In the evaluation of the Diversity Project, it argued that standardization as a universal norm would render useless many samples that have been collected over decades and which can be found in many laboratories (NRC, 1997).

20. The NRC evaluation (1997: 5), however, discouraged a focus on such a list and suggested that a start should be made from the ongoing research practice.

21. See, Fujimura (1987, 1992); Timmermans and Berg (1997); Bowker and Star (1999).

22. After I had finished my research, I worked in a laboratory where such kits were validated and optimized for routine use. For some of the technicians it was a kind of "sport" to experiment with the kits, mainly by diluting the reagents, in order to find their breaking point.
23. Frozen moments is a term used by Haraway (1991a).
24. See Akrich (1997).
25. For the persistence of race and racism in every day life, see Essed (2002).
26. See Star (1991, 1995b); Fujimura (1992, 1995).
27. Star (1995a: 111).
28. Verran (2002: 757).
29. At this point, it serves well to repeat a quotation (from the introductory chapter) about the contribution aimed at by the Diversity Project vis à vis race. Diversity research will be "leading to a greater understanding of the nature of differences between individuals and between human populations . . . and will make a significant contribution to the elimination of racism" (HUGO, 1993: 1).
30. See, for example, Fujumura (1998); Hayden (1998); Lock (2001). Cory Hayden and Joan Fujimura argued that genetics never served as a means to fight against racism and took the statement about race to be merely rhetoric by the Diversity Project. However, to state that genetics never served as a means to fight against racism would be to underestimate the impact and the effect of, for example, the UNESCO Statement on Race, or that of the mtDNA theory of Alan Wilson (the so-called out of Africa theory). Both Wilson's theory of common origin and the UNESCO statement have contributed to broad debates about race and scientific racism. On the lived-in reality of race and racism, see Cross and Keith (1993). In this volume, see especially the contribution of Smith (1993).
31. See Barker (1982); Cross and Keith (1993); Essed and Goldberg (2002).
32. See for large collection of classical papers, Harding (1993).
33. See, for example, the bestseller by Sykes (2001).
34. Strathern (1995b).
35. See, for examples, Jaggar (1983).
36. The Centre for Gender and Diversity at Maastricht University reflects this thinking in its name.
37. See, for examples, Mol (1985); Haraway (1991c,d,e); Oudshoorn (1994); Hirschauer and Mol (1995).
38. Moreover Donna Haraway, Annemarie Mol and Judith Butler have argued that in a "colonizing move," the cultural of gender has become parasitic of the biological/material of sex. They plead for a refiguration (Haraway) of the materiality of sex. See Haraway (1991c, 1997a); Mol (1991a).
39. See especially Hirschauer and Mol (1995), but also Butler (1990) and Chapter 1.
40. See, Strathern (1988: 311–339); Hirschauer and Mol (1995).
41. UNESCO (1951). The issue of debate in this document was not so much the existence of physical differences between races/populations as such. Rather it aimed at questioning the distribution of presumed innate intelligence according to biologically defined clusters of people.
42. UNESCO (1952: 36–91).
43. UNESCO (1952). For an elaboration on both statements, how they are intertwined with feminist politics and for situating some of the scientists involved, see Haraway (1992; particularly pp. 197–206). For an overview of ongoing scientific research on race in the 1950s and 1960s, see Livingstone (1993).

44. See on how culture may be performed as race, Barker (1982).
45. For eighteenth and nineteenth century classifications of the races by, for example, Linnaus, Blumenbach, Lamarck and Cuvier, see Molnar (1975); Gould (1993); Livingstone (1993); Hannaford (1996).
46. See also Strathern (1995c). Developing an argument about knowledge about genetics in relation to kinship and new reproductive technologies, Strathern (1995c: 358) stated: "Perhaps that knowledge will turn what once was a symbol for the immutables of human existence (genetic endowment) into symbol for the open-endedness of possibilities (the realization of potential) that only the individual will manifest."
47. For the kind of troubles produced for scientists, see Haraway (1997a: 239); see also Haraway (1992: 197–203).
48. See, for examples, Stepan (1986); Bloom (1994); Watson-Verran and Turnbull (1995); Anderson (2000, 2002); Verran (2002).
49. A similar argument was developed by Herzig (2000). For exceptions on the materiality of race, see Mol (1991b); Mol and Hendriks (1995).
50. This does not mean, however, that cultural arguments have not been embraced in racist politics and that cultural differences were not essentialized. For examples of how this has been done in recent UK politics, see Barker (1982). Similarly, gender differences were at places essentialized by feminist scholars; for various examples, see Haraway (1991c).
51. For critical examinations of this, see Duster (1992, 1996); Lewontin *et al.* (1993); Essed and Goldberg (2002).
52. For this argument, see Barker (1982); Mol (1990).
53. I am aware that I am neglecting the possessive claim in the term "having"; however let me state this. "To have," based on the possessive right in the self that C. B. Macpherson has wonderfully discerned for Western modernity, enables the agency of the subject. The question is, however, where to locate the action? Now that the subject has been decentered, agency goes well beyond the power of those who have by nature or law, and well beyond the intentionality of delegation. Macpherson (1990: 3) described liberal democratic theory as follows: "Its possessive quality is found in its conception of the individual as essentially the proprietor of his own person or capacities, owing nothing to society for them." The redistribution of power as in the case of DNA leaves little to the expectation and may well question our very concept of possession in capitalist societies; see Marx (1974).
54. See for a critique of the concept of diversity, Hayden (1998).
55. For an elegant working with Haraway's metaphor of vision, see Law (2000).
56. On maps and the worlds they make mobile, see Latour (1990).
57. Strathern (1991: xvii).
58. See for a similar plead, Wade (2002).
59. Lock (2001: 76).
60. See Cann (1998); Cann *et al.* (2002). The list of these cell lines is available online (http://www.cephb.fr/HGDP-CEPH_Table1-1html).
61. Cann et al. (2002: 262).
62. Steve Olson (2001).
63. See Steve Olson (2001).
64. For the Chinese Diversity Project, see, for example, Cavalli-Sforza (1998); Chu *et al.* (1998). For the Indian Diversity Project, see, for example, Bagla (1999); Majumder (2001).

65. On the Genetic Variation Program, see http://www.genome.gov/10001551; http://www.fhcrc.org/pubs/center_news/2000/Oct19/NIH.sht. On the Environmental Genome Project, see Department of Energy (1998); Booker (2001); Greene (2001); Lauerman (2001).
66. On writing technologies and the situatedness of texts, see Haraway (1991a); on the performativity of texts, their effect on subject and object of research as well as the reader, see Mol (2002); Law (2002).
67. Martin (1991, 1999); Keller (1992a). On race and hierarchies, see Lewontin *et al.* (1984); Harding (1993); Hannaford (1996).
68. See Bryld and Lykke (2003).
69. Woolgar and Latour (1986: 64).
70. Examples of these are Knorr-Cetina (1981); Law and Williams (1982); Lynch (1985); Woolgar and Latour (1986); Traweek (1988). For an overview of these studies, see Knorr-Cetina (1983, 1995). For a discussion of the similarities and differences between the ethnomethodological approach and the constructivist approach of the early laboratory studies, see Lynch (1997).
71. Kuhn (1970). The laboratory is also a theoretical notion. A nice twist lies in the fact that a powerful site of Western culture, deeply embedded in society and with ever further-reaching links between domains, is made local and strange through laboratory studies. Instead of science being addressed as the temple of rationality and universality, it is addressed as the specific, that which should be understood in terms of a local culture. Furthermore laboratories are not just seen as a space where the scientist investigates the object but as an organization with an agency in itself, ordering and transforming objects and scientists and making specific alignments between them. In the context of a laboratory, neither the scientist nor the object can be seen as a stable entity. They are linked in specific ways in order to get the job done (Knorr-Cetina, 1995). On the lack of stability of entities (bodies) or, better, on how the various bodies (including that of the surgeon) have to be performed in specific ways in an operating theatre, see Hirschauer (1991).
72. Woolgar and Latour (1986: 64), emphasis in original. See also Ian Hacking (1992).
73. Lynch (1985).
74. Gert-Jan van Ommen, interview with the author on 4th September 1995.
75. See also Lynch *et al.* (1983).
76. Haraway (1997a: 250).
77. In their examination of the symbolic interactionist and the semiotic method, Annemarie Mol and Jessica Mesman (1994) have also traced these different notions of politics. Whereas symbolic interactionism considers the relation between different groups of people, those who can speak and those who are silenced, the semiotic approach to politics focuses on the enactment of different entities in relation to another.
78. For example, in the interviews conducted for this study by the author with Svante Pääbo, 4th February 1997, and with Peter de Knijff, 1st July 1996.
79. *Human Genome Variation in Europe: DNA Markers*, November 1995 in Barcelona.
80. This laboratory has now moved to Leipzig and become part of Max Planck Institute for Evolutionary Anthropology.
81. Meanwhile, the Forensic Laboratory for DNA Research has also grown larger and has taken many more research projects on board.
82. For an articulation of this particular fear of the scientific researcher, namely not to be where the action is, see Law (1994: 45–47).
83. For a similar account, see Star (1995c: 8).

84. See for example, Gupta and Furgusson (1997). On the ordering of both the ethnographic work as well as the ethnography, see Traweek (1992, 1999); Law (1994). For an articulation of the struggle with methodology, see Mol and Mesman (1994).
85. See, for an elegant treatment of this, Geertz (1996).
86. See Clifford and Marcus (1986), particularly James Clifford's contribution in this volume. (pp. 98–121)
87. For an elaboration on this notion, see Strathern (1991).
88. See especially Law (2000, 2002).
89. For this debate in the field of anthropology, see Clifford and Marcus (1986); for examples in the field of STS, see Traweek (1992); Law (1994, 2002); Mol (2002).
90. See also Traweek (1992) and Law (1994). Specifically on the normativity of both author and object of research, and the normativity of the method of research and one's writing, see Mol and Mesman (1994); Law (2002); Mol (2002).
91. See, for various examples, Clarke and Fujimura (1992b); Pickering (1992); Reid and Traweek (2000).
92. This difference between fields is both artificial and real. Despite important overlaps, it seems that both social studies of science and gender studies exist in separate spheres. For example, feminist scholars have been wary of laboratory studies and studies that have science as their main focus. They argued that, now that immense energy had been spent to reveal women in the history of science, especially women as objects of science, studies of laboratories are redirecting attention to domains populated mainly by men. Another reason for the separate spheres is that in the social studies of science, little effort has been made to address gender, let alone racial, issues. However, the opposite can also occur. For many feminist scholars "science seems to be in action" in a relevant sense, when it deals with women, women's lives, sexuality, reproduction and biology, or with female bodies. In contrast to this, Donna Haraway (1997b) in a fascinating lecture wittily told a history of STS from a feminist point of view. The twist was in the very treatment of STS as a branch of feminist studies and not the other way round. In her genealogy, the beginning of STS could be located in the late 1960s and early 1970s, when women started to demand reproductive freedom. She quoted feminist contributions, from the early pamphlets on to the various scholarly works being produced until the late 1990s, and the importance of that for the development of STS as a field.
93. For an example on sexual biases in biology, see Birke (1983); on race, see Gordon, (1976, 1979); Lewontin *et al.* (1984). For an example of gender bias in the language of science, see Keller (1985, 1992a).
94. See Martin (1991); For the distribution of agency to the cytoplasm/germplasm, see Haraway (1992); Stepan and Gilman (1993); Keller (1995a); Graves (2001).
95. See for examples, Keller (1995a); Stamhuis and de Knecht-van-Eekelen (1997).
96. For some examples of these approaches, see Harding (1986, 1993); Bleier (1988). In most cases, however, contributions in this fields resist the rather artificial distinctions (language, sociology, and method) that I try to force on them here. Consider, for example, Evelyn Fox Keller's work, which obviously dealt with language and metaphors in science but hardly ever treats this separately from methods of science.
97. Haraway (1991e: 185), emphasis added.
98. Mol (2002).
99. For a similar argument, see Star (1995c).

100. Within STS, questions of normativity and technology are most persistently articulated by Donna Haraway, John Law, Annemarie Mol and Susan Leigh Star.
101. For collections of studies from this angle, see Law (1991, 1999); Star (1995b); Berg and Mol (1999); Lock and Cambrosio (2000).
102. See, for examples, Franklin and Ragoné (1998); Rapp (2000); Mol (2002).
103. For a very elegant example that deals with normalization and naturalization, see Cussins (1998); on standardization and marginalization see the classic paper by Star (1991); for a beautiful example on personhood, see Moser and Law (1999); on performing the sexes and its articulation in technologies, see Hirschauer and Mol (1995); on how masculinity and Euro-American concepts of kinship are implicated in computer simulation, see Helmreich (1998); on sex and gender, see also Hirschauer (1998); on the body and its articulation in instruments and technologies, see Hirschauer (1991); Oudshoorn (1994); Mol (2002).

Glossary

Adenine (A) One of the four main building blocks of DNA. Adenine is a nitrogenous base and one member of the base pair A–T (adenine–thymine).

ALF sequencing automated laser fluorescent sequencing, which detects target DNA fragments because they carry a fluorescent label with them. The ALF$^{®}$ machine (Pharmacia) comes with a software program that enables its user to analyze and reorder the data.

Allele One of several alternative forms of a gene occupying a given locus on the chromosome, or a variable locus in non-coding DNA. A single allele for each locus is inherited separately from each parent, so every individual has two alleles for each gene. The term allele is also used for a variable fragment of mitochondrial DNA.

Allele frequency Relative proportions of all varying alleles of a specific DNA locus in a population.

Autosome A chromosome that is not involved in sex determination. The diploid human genome consists of 46 chromosomes: 22 pairs of autosomes and 1 pair of sex chromosomes (X- and Y-chromosome). Each parent contributes one haploid set of 22 autosomal chromosomes and 1 sex chromosome to each offspring.

Back-cross Crossing an organism with one of its parents or with a genetically equivalent organism. The offsprings of such a cross are referred to as the back-cross generation or back-cross progeny.

Base The nitrogen-containing components of the nucleic acids. There are four bases in DNA: adenine (A), guanine (G), cytosine (C) and thymine (T). The sequence of bases in coding DNA determines the genetic code. Three bases for each amino acid in a protein.

Base-pair Two bases – adenine and thymine, guanine and cytosine – held together by weak bonds. In double-stranded nucleic acids, one base of each pair occurs on each strand. Two strands of DNA are thus held together by the bonds between these base-pairs.

Bottleneck, genetic a reduction in genetic diversity, for example as a result of a reduction in population size.

Cell The basic structural building block of living organisms. Some organisms consist of a single cell (e.g. bacteria). Multicellular organisms contain eukaryotic cells, which contain a nucleus, enclosing DNA, and cytoplasm surrounded by a membrane. Cells can be classified as germline (sperm, egg) or as somatic (rest of body tissues).

Cell line   A culture of cells that can be kept alive indefinitely through the appropriate supply of nutrients. The first human cell line ever produced was the HeLa cell line (1952), derived from the African-American woman Henrietta Lacks.

Chromosome   A rod-like structure consisting of chromatin and carrying genetic information arranged in a linear sequence. The backbone of the chromosome is a very long molecule of DNA. In humans, all somatic cells contain 23 pairs of chromosomes (44 autosomes plus the two sex chromosomes). Non-somatic cells, the gametes (the egg cell or the sperm), contain half the number of the chromosomes (22 autosomes and one sex chromosome).

Code   The sequence of DNA bases which forms the instructions for a given characteristic or trait. The triplet pattern of the DNA bases serves to specify the order of amino acids in proteins. For example, the triplet GAA codes for the amino acid glutamine.

Coding region   The part of a DNA sequence that codes for the amino acids in a protein.

Common ancestor   Any ancestor of both one's father and one's mother. *See* MRCA (most recent common ancestor).

Complementary sequences   Nucleic acid base sequences that can form a double-stranded structure of DNA by matching base-pairs; the complementary sequence to GTAC is CATG.

Control region   A hypervariable region of the non-coding part of the sequence located on mitochondrial DNA, consisting of the D-loop and flanking regions. The control region is clustered as hypervariable region I (HVI) and hypervariable region II (HVII).

Cytoplasm   Part of the cell outside the nucleus, contained within its plasma membrane.

Cytosine (C)   One of the four main building blocks of DNA. Cytosine is a nitrogenous base and one member of the base pair G–C (guanine–cytosine).

Denaturation   Loss of normal three-dimensional shape of a macromolecule, such as DNA, without breaking covalent bonds, usually accompanied by loss of its biological activity. In the case of DNA, denaturation is usually the conversion of DNA from the double-stranded into single-stranded form.

Diploid   A full set of paired chromosomes: one chromosome from each parent. Most animal cells except the gametes have a diploid set of chromosomes. The diploid human genome has 46 chromosomes. *See* haploid.

D-loop   The displacement loop is a hypervariable region of the non-coding part of the sequence located on mitochondrial DNA.

DNA (deoxyribonucleic acid)   The molecule that encodes genetic information. DNA is a double-stranded chain of nucleotides held together by weak bonds between base pairs.

DNA fingerprinting   The compilation of a DNA profile of one individual in such a way that it can be used for identification. Different highly variable regions of the DNA (genetic markers) are investigated and the specificity of these variable regions in one individual's DNA produce together a DNA profile.

DNA profile   *See* DNA fingerprinting.

DNA sequence   The relative order of base-pairs, whether in a fragment of DNA, a gene, a chromosome or an entire genome.

DNA typing   *See* DNA fingerprinting.

Double helix   The natural shape of DNA; the coiled conformation of two complementary, anti-parallel chains of nucleotides.

EDTA   Ethylenediaminetetraacetic acid.

Electrophoresis   A method for separating fragments of DNA from a mixture containing fragments of different size. An electric current is passed through a medium containing the mixture, and each DNA fragment travels through a medium at a different rate, depending on its electrical charge and size. The fragments are, therefore, separated according to their size.

ELSI   Ethical, legal and social implications of the Human Genome Project.

Enzyme   Protein molecule that acts as a catalyst, speeding up the rate at which a biochemical reaction proceeds without altering the direction or nature of the reaction.

Eukaryotic cell   A cell with a distinct nucleus.

Gamete   Mature male or female reproductive cell (sperm or ovum) with a haploid set of chromosomes (23 for humans).

Gene   The fundamental physical and functional unit of heredity. A gene is an ordered sequence of nucleotides located in a particular position on a particular chromosome. Each gene encodes a specific functional product, such as a protein. *See* also allele.

Genetic code   The triplets of nucleotides. The four letters of the DNA alphabet form 64 triplets, or codons, which specify the 20 different amino acid subunits and the stop signals that end the production of the protein. Hence, most amino acids are coded by more then one triplet.

Genetic mapping   Determination of the relative position of genes or genetic markers on a DNA molecule and of the distance between them.

Genetic marker   An identifiable physical location on a chromosome whose inheritance can be monitored. Markers can be expressed regions of DNA (genes), a sequence of bases that can be identified by restriction enzymes, or a segment of DNA with no known coding function.

Genome   All the DNA contained in an organism or a cell, which includes both the chromosomes within the nucleus and the DNA in mitochondria.

Genotype   The total genetic or hereditary constitution that an individual receives from his or her parents.

Guanine (G)   One of the four main building blocks of DNA. Guanine is a nitrogenous base and one member of the base pair G–C (guanine–cytosine).

Haploid   A single set of chromosomes (half the full set of genetic material) present in the egg and sperm cells of animals. The reproductive cells of humans contain 23 chromosomes (*see* diploid).

HGDP   Human Genome Diversity Project (referred to in this book as the Diversity Project). The global initiative to map and sequence human genome diversity.

HGP   Human Genome Project.

HUGO   The Human Genome Organization is the political organization of the HGP; it is the international organization of scientists involved in the HGP. It was established in 1989 by a group of genome scientists to promote international collaboration within the project.

Ladder   In electrophoresis (a method of separating fragments of DNA), a ladder is an extra sample of known size that helps to determine the size of the target fragment. Usually, the ladder consists of all possible size variations that could be found in such a fragment. Some ladders are called *universal ladders*. They do not contain the variable fragment sizes but rather fixed measures such as 50, 100, 150 base-pairs, etc. These ladders are also called sizers, and serve as molecular weight "markers".

Lineage An ancestor–descendant sequence of populations, cells or genes of which the pattern of inheritance can be distinguished from other sequences. Also, a line of common descent.

Locus The position on a chromosome, a gene or other chromosomal markers; also the DNA sequence at that position.

Marker *See* genetic marker.

Metabolism The sum total of the various biochemical reactions occurring in a living cell, required for growth and the maintenance of life.

Meiosis The process of two consecutive cell divisions in the diploid progenitors of sex cells. Meiosis results in four rather than two daughter cells, each with a haploid set of chromosomes.

Mitochondria (one mitochondrion) Membrane-bounded organelles found outside the nucleus in the cytoplasm of all human cells and which carries a specific type of DNA (mtDNA). Mitochondria are often described as the powerhouse of the cell, since they provide the cell with energy.

Mitochondrial DNA (mtDNA) A small circular genome located in mitochondria. This DNA is inherited via the mother only since it is transmitted via the egg cell. The cytoplasm of a sperm cell does not enter the egg cell and hence the paternal mtDNA is not found in the fertilized egg cell.

Molecular clock The postulation that the nucleotide substitutions, or other changes in the DNA, occur at a constant rate for a given family of genes, or of genetic markers. The hypothesis is that DNA changes are linear with time, and constant over different populations and in different places. If that is so, then the sequence differences found in different populations can be used to estimate time since divergence.

MRCA (most recent common ancestor) A term used to describe the possible temporal and ancestral relationship between alleles. It is applied to a locus-by-locus comparison, not to groups of individuals.

Mutation A heritable change in DNA sequence.

Mutation rate The probability of a new mutation in a particular gene or a fragment of non-coding DNA, either per gamete or per generation.

NIH US National Institutes of Health.

NRC US National Research Council.

Nuclear DNA The DNA-containing chromosomes located in the nucleus (46 in humans).

Nucleic acid A large molecule composed of nucleotide subunits. *See also* DNA.

Nucleotide A building block of DNA containing a base.

Nucleus Membrane-bounded organelle in eukaryotic cells that contains the chromosomes.

Organelle Membrane-bounded particles (e.g. mitochondria, chloroplasts) in eukaryotic cells that contain enzymes for specialized functions.

PCR *See* polymerase chain reaction.

Polymerase chain reaction (PCR) A laboratory method for the exponential amplification of a selected fragment of DNA. The procedure requires two kinds of synthesized primers (essential to initiate DNA synthesis), each kind complementary to just one end of the target DNA fragment; a thermostable DNA polymerase; and a supply of nucleotides. First, a solution containing the DNA fragment, the primers and the nucleotides is heated, and the two strands of DNA come apart; the primers then anneal to the appropriate ends. After the solution is cooled, the polymerase is added, and the enzyme effects the replication of the DNA fragment between the two primers on the ends by using the nucleotides in the solution. Each newly synthesized strand of DNA subsequently serves as a template for yet another strand, so the supply doubles with each repetition of the procedure. The use of PCR enables the genetic analysis of biological samples containing only tiny amounts of DNA.

Polymerase, DNA   An enzyme that acts as a catalyst in the replication of DNA. *See also* Taq polymerase.

Polymorphism   Difference in DNA sequence among individuals. Genetic variations occurring in more than 1% of a population would be considered useful polymorphisms for genetic linkage analysis.

Population genetics   Application of Mendel's laws and other principles of genetics to entire populations of organisms.

Population substructure   Organization of a population into smaller breeding groups between which migration is restricted. Also called population subdivision.

Primary structure   The sequence order along the backbone of a macromolecule, such as DNA, composed of different building blocks. *See also* secondary and tertiary structure.

Primer   Short pre-existing polynucleotide chain to which new nucleotides can be added by DNA polymerase. *See also* polymerase chain reaction (PCR).

Protein   A large molecule composed of one or more chains of amino acids in a specific order. This order is determined by the sequences of nucleotides in the gene coding for the protein. Proteins are required for the structure, function and regulation of cells, tissue and organs; each protein has unique functions.

RAFI   Rural Advancement Foundation International.

Random mating   System of mating in which mating pairs are formed independently of genotype and phenotype.

Recombinant DNA technologies   Procedures used to join together DNA segments in a cell-free system (an environment outside a cell organism). Under appropriate conditions, a recombinant DNA molecule can enter a cell and replicate there, either autonomously or after it has become integrated into a cellular chromosome.

Recombination, chromosomal   The interchange of chromosomal segments (by breaking and rejoining) between a pair of (homologue) chromosomes during replication. Since a piece of one chromosome now resides on the other chromosome in the pair, and vice versa, the recombination is said to be heterologous. Also called "crossing over."

Replication, DNA   The use of existing DNA as a template for the synthesis of new DNA strands. In eukaryotes, replication occurs in the cell nucleus before cell division.

Secondary structure   The folding pattern of the DNA in regular structures such as spirals (α-helix).

Sequence, DNA   See DNA sequence.

Sequencing   Determination of the order of nucleotides in a DNA.

Sex chromosomes   The X- and Y-chromosomes in human beings that determine the sex of an individual. Females have two X-chromosomes in diploid cells; males have an X- and a Y-chromosome. Reproductive cells contain only one sex chromosome.

Short tandem repeat (STR)   Multiple copies of a short nucleotide sequence (tandem) in a particular DNA fragment (such as, ACTACTACTACTACT). They may be highly variable and are used as markers in population studies and in forensic DNA evidence.

Sizer   *See* ladder.

Taq polymerase   Thermostable polymerase enzyme (*Thermus aquaticus* polymerase). Since it is thermostable, this enzyme does not lose its biological function even at high temperatures (e.g. 92 °C). It is, therefore, an important component of the polymerase chain reaction (q.v.) for the amplification of DNA fragments. *See* polymerase.

Template DNA   DNA strand that is transcribed and so determines the sequence of its (mirror) copy. In the laboratory context, this is the DNA material that is used for replication in the polymerase chain reaction.

Tertiary structure   Complex three-dimensional structure of a macromolecule, such as DNA.

UNESCO   United Nations Educational, Scientific and Cultural Organization.

Universal ladder   *See* ladder.

Thymine (T)   One of the four main building blocks of DNA. Thymine is a nitrogenous base and one member of the base pair A–T (adenine–thymine).

Transcription   The process of producing RNA (ribonucleic acid) from coding DNA, as part of producing proteins.

References

Akrich, M. (1997). The de-scription of technical objects. In *Shaping Technology/ Building Society: Studies in Sociotechnical Change*, 2nd edn, Bijker, W. and Law, J. (eds.). Cambridge, MA: MIT Press, pp. 205–225.

Alper, B., Wiegand, P. and Brinkmann, B. (1995a). Frequency profiles of 3 STRs in a Turkish Population. *International Journal for Legal Medicine* **108**: 110–112.

Alper, B., Meyer, E., Schurenkamp, M. and Brinkmann, B. (1995b). HumFES/FPS and HumF13B: Turkish and German population data. *International Journal for Legal Medicine* **108**: 93–95.

Amann, K. and Knorr-Cetina K. (1990). The fixation of (visual) evidence. In *Representation in Scientific Practice*, Woolgar, S. and Lynch, M. (eds.). Cambridge, MA: MIT Press, pp. 85–121.

Anderson, S., Bankier, A. T., Barrell, B. G. *et al.* (1981). Sequence and organization of the human mitochondrial genome. *Nature* **290**: 457–465.

Anderson, W. (2000). The possession of Kuru: medical science and biocolonial exchange. *Society for Comparative Study of Science and History* **10**: 713–744.

(2002). Postcolonial technoscience. *Social Studies of Science* **32**: 643–658.

Andrews, R. M., Kubacka, I., Chinnery, P. F. *et al.* (1999). Reanalysis and revision of the Cambridge reference sequence for human mitochondrial DNA. *Nature Genetics* **2**: 147.

Ankel-Simons, F. and Cummins, J. M. (1996). Misconceptions about mitochondria and mammalian fertilization: implications for theories on human evolution. *Proceedings of the National Academy of Science USA* **93**: 1359–1363.

Annas, G. J. (2001). Genism, racism, and the prospect of genetic genocide. In *UNESCO 21st Century Talks: The New Aspects of Racism in the Age of Globalization and the Gene Revolution*, Durban, September. http://www.bumc.bu.edu/www/sph/lw/pvl/genism.htm.

Anon. (1995). Bias-free interracial comparison. [Editorial.] *Nature* **377**: 183–184.

(2000). Census, race and science. [Editorial] *Nature Genetics* **24**: 97–98.

(2001a). The human genome. *Nature* **409**: special issue.

(2001b). The human genome. *Science* **291**: special issue.

Arnason, E. (2003). Genetic heterogeneity of Icelanders. *Annals of Human Genetics* **67**: 5–16.

192

Austin, J. L. (1962). *How to do Things with Words*. Cambridge, MA: Harvard University Press.

Bagla, P. (1999). *Genomics: India Prepares Research, Policy Initiative*. http://www.gene.ck/info4action/1999/Jan/msg00077.html.

Barker, M. (1982). *The New Racism*. London: Junction Books.

Berg, M. and Mol, A. (eds.) (1999). *Differences in Medicine: Unravelling Practices, Techniques and Bodies*. Durham NC: Duke University Press.

Bertranpetit, J. (ed.) (1995). *Euroconference on Human Genome Variation in Europe: DNA Markers, November*, Barcelona, November.

(1996) European Human Genome Diversity Project Regional Committee Meeting, London, January.

Birke, L. (1983). *Women, Feminism and Biology*. Brighton, UK: Wheatsheaf Books, Harvester Press.

Bleier, R. (ed.) (1988). *Feminist Approach to Science*. New York: Pergamon Press.

Bloom, L. (1994). Constructing whiteness: popular science and *National Geographic* in the age of multiculturalism. *Configurations* **2**: 15–33.

Booker, S. M. (2001). *Environmental Genome Project: A Positive Sequence of Events*. *NIEHS News*. http://ehpnet1.niehs.nih.gov/docs/2001/109-1/niehsnews.html.

Börner, G. V., Yokobori, S., Morl, M., Dorner, M. and Paabo, S. (1997). RNA editing in metazoan mitochondria: staying fit without sex. *FEBS Letters* **409**: 320–324.

Borst, P. and Grivell, L. A. (1981). Small is beautiful: portrait of a mitochondrial genome. *Nature* **290**: 443–444.

Bosk, C. L. (1992). *All God's Mistakes: Genetic Counseling in a Pediatric Hospital*. Chicago, IL: University of Chicago Press.

Bouquet, M. (1995). Exhibiting knowledge: the tree of Dubois, Haeckel, Jesse and Rivers at the *Pithecanthropus* Centennial Exhibition. In *Shifting Contexts: Transformations in Anthropological Knowledge*, Strathern, M. (ed.). London: Routledge, pp. 31–55.

Bowker, G. and Star, S. L. (1999). *Sorting Things Out*. Cambridge, MA: MIT Press.

Bromham, L., Eyre-Walker, A., Smith, N. H. and Smith, J. M. (2003). Mitochondrial Steve: paternal inheritance of mitochondria in humans. *Trends in Ecology and Evolution* **18**: 2–4.

Bryld, M. and Lykke, N. (2003). Van Rambosperma naar Konigin Eicel: Twee Filmversies van de Menselijke Voortplanting van Wetenschapsfotograaf Lennart Nilson. *Tijdschrift voor Genderstudies* **1**: 30–44.

Burckhardt, F., von Haeseler, A. and Meyer, S. (1999). HvrBase: compilation of mtDNA control region sequences from primates. *Nucleic Acids Research* **27**: 138–142.

Butler, D. (1995). Genetic diversity proposal fails to impress international ethics panel. *Nature* **377**: 373.

Butler, J. (1990). *Gender Trouble: Feminism and the Subversion of Identity*. London: Routledge.

(1993). *Bodies that matter: On the Discursive Limits of "Sex"*. London: Routledge.

Callon, M. (1986a). Some elements of a sociology of translation: domestication of the scallops and the fishermen of St. Brieuc Bay. In *Power, Action and Belief: A New Sociology of Knowledge?* Law, J. (ed.). London: Routledge & Kegan Paul, pp. 196–233.

(1986b). The sociology of an actor-network: the case of the electric vehicle. In *Mapping the Dynamics of Science and Technology: Sociology of Science in the Real World,* Callon, M., Law, J. and Rip, A. (eds.). Basingstoke: Macmillan, pp. 19–34.

(1991). Techno-economic networks and irreversibility. In *A Sociology of Monsters: Essays on Power, Technology and Domination*, Law, J. (ed.). London: Routledge, pp. 132–165.

Cann, H. M. (1998). Human genome diversity. *Académie des Science* **321**: 443–446.

Cann, H. M., de Tomas, C., Cazes, L. *et al*. (2002). A human genome diversity cell line panel. *Science* **296**: 261–262.

Cann, R. L., Stoneking, M. and Wilson, A. C. (1987). Mitochondrial DNA and human evolution. *Nature* **325**: 31–36.

Castañeda, C. (1998). Heredity in science and medicine. In *EASST Annual Meeting*, Lisbon.

Cavalli-Sforza, L. L. (1991). Genes, peoples and languages. *Scientific American* **265**: 72–78.

(1993). *Answers to Frequently Asked Questions About the Human Genome Diversity Project.* Stanford: The North American Committee.

(1995). The Human Genome Diversity Project. In *Actes 1995: Proceedings of the International Bioethics Committee of UNESCO*, Vol. 2, pp. 71–83.

(1998). The Chinese Human Genome Diversity Project. *Proceedings of the National Academy of Science USA* **95**: 11501–11503.

Cavalli-Sforza, L. L., Wilson, A. C., Cantor, C. R., Cook-Deegan, R. M. and King, M.-C. (1991). Call for a worldwide survey of human genetic diversity: a vanishing opportunity for the Human Genome Project. *Genomics* **11**: 490–491.

Cavalli-Sforza, L. L., Menozzi, P. and Piazza, A. (1994). *The History and Geography of Human Genes*. Princeton, NJ: Princeton University Press.

Chakraborty, R. and Kidd, K. (1991). The utility of DNA typing in forensic work. *Science* **254**: 1735–1739.

Chakraborty, R., Srinivasan, M. R. and Daiger, S. P. (1993). Evaluation of standard error and confidence interval of estimated multilocus genotype probabilities, and their implications in DNA forensics. *American Journal for Human Genetics* **52**: 60–70.

Chapman, M. (ed.) (1993). *Social and Biological Aspects of Ethnicity*. Oxford: Oxford University Press.

Chu, J. Y., Huang, W., Kuang, S. Q., Wang, J. M. *et al*. (1998). Genetic relationship of populations in China. *Proceedings of the National Academy of Science USA* **95**: 11763–11768.

Clarke, A. E. (1995). Research materials and reproductive science in the United States, 1910–1940. In *Ecologies of Knowledge: Work and Politics in Science and Technology*, Star, S. L. (ed.). New York: State University of New York Press, pp. 183–226.

Clarke, A. E. and Fujimura, J. H. (1992a). What tool? Which job? Why right? In *The Right Tool for the Job: At Work in Twentieth-Century Life Science*, Clarke, A. and Fujimura, J. (eds.). Princeton, NJ: Princeton University Press, pp. 3–43.

(eds.) (1992b). *The Right Tool for the Job: At Work in Twentieth-Century Life Science*. Princeton, NJ: Princeton University Press.

Clifford, J. (1986). On ethnographic allegory. In *Writing Culture: The Poetics and Politics of Ethnography*, Clifford, J. and Marcus, G. (eds.). Berkeley, CA: University of California Press, pp. 98–121.

Clifford, J. and Marcus, G. (eds.) (1986). *Writing Culture: The Poetics and Politics of Ethnography*. Berkeley, LA: University of California Press.

Cole, S. A. (1998). Witnessing identification: latent fingerprinting evidence and expert knowledge. *Social Studies of Science* **28**: 687–712.

Cross, M. and Keith, M. (eds.) (1993). *Racism, the City, and the State*. New York: Routledge.

Cunningham-Burly, S. and Kerr, A. (1999). Defining the "social": towards and understanding of scientific and medical discourses on the social aspects of the new human genetics. *Sociology of Health and Illness* **21**: 647–668.

Cussins, C. (1998). Producing reproduction: techniques of normalization and standardization in infertility clinics. In *Reproducing Reproduction: Kinship, Power and Technological Innovation*, Franklin, S. and Ragoné, H. (eds.) Philadelphia, PA: University of Pennsylvania Press, pp. 66–101.

Darlington, C. D. (1947). The genetic component of language. *Heredity* **1**: 269–286.

Daston, L. (2000). The coming into beings of scientific objects. In *Biographies of Scientific Objects*, Daston, L. (ed.). Chicago, IL: The University of Chicago Press, pp. 1–15.

de Bont, A. (2000). De Organisatie van een Virus: Hoe een Technocratische WHO Transnationale Kennispolitiek Bedrijft. Ph.D. Thesis, University of Maastricht.

de Knijff, P. (1992). Genetic heterogeneity of apolipoprotein E and its influence on lipoprotein metabolism. Ph.D. Thesis, Leiden University.

de Knijff, P., Kayser, M., Krawczak, M. *et al.* (1995). Y-typing using STR. In *Conference on Human Genome Variation in Europe: DNA Markers*, Barcelona, p. 31.

de Knijff, P., Kayser, M., Caglia, A. *et al.* (1997). Chromosome Y microsatellites: population genetic and evolutionary aspects. *International Journal of Legal Medicine* **110**: 134–149.

de Stefano, P. (1996). Genomics 101: the X's and Y's of legal rights to genetic material. *IP-Worldwide: The Magazine of Law and Policy for High Technology*, vol. 101. Online http://ipmag.com/destefan.html.

Dennis, C. (2003). Error reports threaten to unravel databases of mitochondrial DNA. *Nature* **421**: 773–774.

Department of Energy (1992). *DOE Human Genome Project: Primer on Molecular Genetics*. Washington, DC: US Department of Energy.

    (1998). New HGP spinoff programs to study genes for environmental risk. *Human Genome News* **9**: 1–2.

Dickson, D. (1996). Whose genes are they anyway? *Nature* **381**: 11–14.

Dorit, R. L., Akashi, H. and Gilbert, W. (1995). Absence of polymorphism at the ZFY locus on the human Y-chromosome. *Science* **268**: 1183–1185.

Drouin, J. (1980). Cloning of human mitochondrial DNA in *Escherichia coli. Journal of Molecular Biology* **140**: 15–34.

Dunn, L. C. (1951). *Race and Biology*. Paris: UNESCO.

Duster, T. (1992). Genetics, race, and crime: recurring seduction to a false precision. In *DNA on Trial: Genetic Identification and Criminal Justice*, Billings, P. R. (ed.). New York: Cold Spring Harbor Laboratory Press, pp. 129–141.

    (1996). The prism of heredity and the sociology of knowledge. In *Naked Science: Anthropological Inquiry into Boundaries, Power, and Knowledge*, Nader, L. (ed.). New York: Routledge, pp. 119–130.

Essed, P. (2002). Everyday racism: a new approach to the study of racism. In *Race Critical Theories*, Essed, P. and Goldberg, D. T. (eds.). Malde, MA: Blackwell, pp. 176–194.

Essed, P. and Goldberg, D. T. (eds.) (2002). *Race Critical Theories*. Malde, MA: Blackwell.

Franklin, S. and Ragoné, H. (eds.) (1998). *Reproducing Reproduction: Kinship, Power, and Technological Innovation*. Philadelphia, PA: University of Pennsylvania Press.

Fujimura, J. H. (1987). Constructing "do-able" problems in cancer research: articulating alignment. *Social Studies of Science* **17**: 257–293.

(1992). Crafting science: standardized packages, boundary objects, and "translation." In *Science as Practice and Culture*, Pickering, A. (ed.). Chicago, IL: University of Chicago Press, pp. 168–211.

(1995). Ecologies of action: recombining genes, molecularizing cancer, and transforming biology. In *Ecologies of Knowledge: Work and Politics in Science and Technology*, Star, S. L. (ed.). New York: State University of New York Press, pp. 302–347.

(1998). Creating "cultures" in debates about genomes, information, and diversity. In *Proceedings of the Conference on Postgenomics? Historical, Techno-epistemic and Cultural Aspects of Genome Projects*, Berlin, July.

Fujimura, J. and Fortun, M. (1996). Constructing knowledge across social worlds: the case of DNA sequencing databases in molecular biology. In *Naked Science: Anthropological Inquiry into Boundaries, Power, and Knowledge*, Nader, L. (ed.). New York: Routledge, pp. 160–173.

Galison, P. (1999). Trading zone: coordinating action and belief. In *The Science Studies Reader*, Biagioli, M. (ed.). New York: Routledge, pp. 137–161.

Galton, F. (1892). *Finger Prints*. London: Macmillan.

Garfinkel, H. (1996a). Passing and the managed achievement of sex status in an "intersexed" person Part 1. In *Studies of Ethnomethodology*, Garfinkel, H. (ed.). Cambridge: Polity Press, pp. 116–185.

(1996b). *Studies of Ethnomethodology*. Cambridge: Polity Press.

Geertz, C. (1996). *Works and Lives: The Anthropologist as Author*, 2nd edn. Cambridge: Polity Press.

Gilbert, W. (1992). A vision of the grail. In *The Code of Codes: Scientific and Social Issues in the Human Genome Project*, Kevles, D. J. and Hood, L. (eds.). Cambridge, MA: Harvard University Press, pp. 83–98.

Ginsburg, F. D. and Rapp, R. (eds.) (1995). *Conceiving the New World Order: The Global Politics of Reproduction*. Berkeley, CA: University of California Press.

Goffman, E. (1961). *Encounters: Two Studies in the Sociology of Interaction*. Harmondsworth: Penguin University Books.

Goldman, E. (1970). The traffic in women. In *The Traffic in Women and Other Essays on Feminism*, Goldman, E. (ed.). New York: Time Change Press, pp. 19–32.

Gordon, L. (1976). *Woman's Body, Woman's Rights: A Social History of Birth Control in America*. New York: Viking.

(1979). De Strijd voor Vrijheid van Reproduktie: Drie Stadia in het Feminisme. In *Socialisties Feministiese Teksten*. Nijmegen: Feministische Uitgeverij Sara, pp. 36–72.

Gould, S. J. (1993). America polygeny and craniometry before Darwin: Blacks and Indian as separate, inferior species. In *The Racial Economy of Science: Towards a Democratic Future*, Harding, S. (ed.). Bloomington, IA: Indiana University Press, pp. 84–116.

Graves Jr, J. L. (2001). *The Emperor's New Clothes: Biological Theories of Race at the Millenium.* New Brunswick, NJ: Rytgers University Press.

Greene, L. A. (2001). *New Center a Stroke of Gene-ius. NIEHS News.* http://ehpnet1.niehs.nih.gov/docs/2001/109-1/niehsnews.html.

Gupta, A. and Furgusson, J. (1997). Discipline and practice: "the field" as a site, method, and location in anthropology. In *Anthropological Locations*, Gupta, A. and Furgusson, J. (eds.). Berkeley, CA: University of California Press, pp. 1–46.

Gyllensten, U., Wharton, D. and Wilson, A. (1985). Maternal inheritance of mitochondrial DNA during backcrossing of two species of mice. *Journal of Heredity* **76**: 321–324.

Gyllensten, U., Wharton, D., Josefsson, A. and Wilson, A. (1991). Paternal inheritance of mitochondrial DNA in mice. *Nature* **353**: 255–257.

Hacker, S. (1989). *Pleasure, Power, and Technology: Some Tales of Gender, Engineering and the Co-operative Workplace.* London: Unwin Hyman.

Hacking, I. (1992). The self-vindication of laboratory science. In *Science as Practice and Culture*, Pickering, A. (ed.). Chicago, IL: University of Chicago Press. pp. 29–65.

Hagelberg, E., Goldman, N., Lio, P. *et al.* (1999). Evidence for mitochondrial DNA recombination in a human population of island Melanesia. *Proceedings of the Royal Society of London* **266**: 485–492.

Hagendijk, R. (1996). Wetenschap, Constructivisme en Cultuur. Ph.D. Thesis, University of Amsterdam.

Handt, O., Meyer, S. and von Haeseler, A. (1998). Compilation of human mtDNA control region sequences. *Nucleic Acids Research* **26**: 126–129.

Hannaford, I. (1996). *Race: The History of an Idea.* Baltimore, MD: Johns Hopkins University Press.

Haraway, D. J. (1988). Primatology is politics by other means. In *Feminist Approach to Science*, Bleir, R. (ed.). New York: Pergamon Press, pp. 77–118.

(1991a). A cyborg manifesto: Science, technology, and socialist-feminism in the late twentieth century. In *Simians, Cyborgs, and Women: The Reinvention of Nature*, Haraway, D. J. (ed.). London: Free Association Books, pp. 149–181.

(1991b). *Simians, Cyborgs, and Women: The Reinvention of Nature.* London: Free Association Books.

(1991c). Gender for a Marxist dictionary: The sexual politics of a word. In *Simians, Cyborgs, and Women: The Reinvention of Nature*, Haraway, D. J. (ed.). London: Free Association Books, pp. 127–148.

(1991d). In the beginning was the word: The genesis of biological theory. In *Simians, Cyborgs, and Women: The Reinvention of Nature*, Haraway, D. J. (ed.). London: Free Association Books, pp. 71–81.

(1991e). Situated knowledges: the science question in feminism and the privilege of partial perspective. In *Simians, Cyborgs, and Women: The Reinvention of Nature*, Haraway, D. J. (ed.). London: Free Association Books, pp. 183–201.

(1992). *Primate Visions: Gender, Race, and Nature in the World of Modern Science*, 2nd edn. London: Verso.

(1997a). *Modest_Witness@Second_Millennium.FemaleMan©_Meets_OncoMouse™*. New York: Routledge.

(1997b). Feminist science studies: a history of STS from a feminist point of view. In *Proceedings of the WTMC Summel School*, Enschede, September.

Harding, J. (ed.) (1986). *Perspectives on Gender and Science*. London: Falmer Press.

(1993). *The Racial Economy of Science: Towards a Democratic Future*. Bloomington, Indianapolis, IA: Indiana University Press.

Hartl, D. L. (1995). *Essential Genetics*. Sudbury, UK: Jones and Bartlett.

Hasegawa, M., Di Rienzo, A., Kocher, T. D. and Wilson, A. C. (1993). Towards a more accurate time scale for the human mitochondrial DNA tree. *Journal of Molecular Evolution* **37**: 347–354.

Hawks, J., Oh, S., Hunley, K., *et al.* (2000). An Australasian test of the recent African origin theory using the WLH-50 calvarium. *Journal for Human Evolution* **39**: 1–22.

Hayden, C. (1998). A Biodiversity Sampler for the Millennium. In Reproducing Reproduction: Kinship, Power and Technological Innovation, Franklin S. and Ragoné H. (eds.). Philadelphia, PA: Pennsylvania Press, pp. 173–206.

Helmreich, S. (1998). Replicating reproduction: or, the essence of life in the age of virtual electronic reproduction. In *Reproducing Reproduction: Kinship, Power and Technological Innovation*, Franklin, S. and Ragoné, H. (eds.). Philadelphia, PA: Pennsylvania Press, pp. 207–234.

Herrnstadt, C., Preston, G., Andrews, R. *et al.* (2002). A high frequency of mtDNA polymorphisms in HeLa cell sublines. *Mutation Research* **501**: 19–28.

Herzig, R. (2000). Producing "race" in early twentieth-century America. In *Proceedings of the IISG Conference*, Amsterdam.

Hirschauer, S. (1991). The manufacture of bodies in surgery. *Social Studies of Science* **21**: 279–319.

(1998). Performing sexes and genders in medical practices. In *Differences in Medicine: Unravelling Practices, Techniques and Bodies*, Berg, M. and Mol, A. (eds.). Durham, NC: Duke University Press, pp. 13–28.

Hirschauer, S. and Mol, A. (1995). Shifting sexes, moving stories: feminist/constructivist dialogues. *Science, Technology, and Human Value* **20**: 368–385.

Holland, L. (1995). *The Gene Hunters*. Zed Productions.

Hopgood, R., Sullivan, K. M. and Gill, P. (1992). Strategies for automated sequencing of human mitochondrial DNA directly from PCR products. *Biotechniques* **13**: 82–92.

Howell, N., McCullough, D. A., Kubacka, I., Halvorson, S. and Mackey, D. (1992). The sequence of human mtDNA: the question of errors versus polymorphisms. *American Journal for Human Genetics* **50**: 1333–1337.

HUGO (1993). *The Human Genome Diversity (HGD) Project: Summary Document*. Sardinia: Human Genome Organization.

Jacobs, L. (ed.) (1995). *Gerechtelijke Laboratoria in Beeld: Een Kennismaking met Beoefende Deskundigen*. Groningen: Wolters-Noordhof.

Jaggar, A. M. (1983). *Feminist Politics and Human Nature*. Sussex, NJ: Harvester Press, Rowman & Allanheld.

Jasanoff, S. (1995). *Science at the Bar: Law, Science, and Technology in America*. Cambridge, MA: Harvard University Press.

Jasanoff, S. and Lynch, M. (eds.) (1998). Special issue. *Social Studies of Science* Dec.

Jeffreys, A. J., Turner, M. and Debenham, P. (1991). The efficiency of multilocus DNA fingerprint probes for individualisation and establishment of family relationships, determined from extensive casework. *American Journal for Human Genetics* **48**: 824–840.

Jobling, M. A. and Tyler-Smith, C. (1995). Fathers and sons: the Y chromosome and human evolution. *Trends in Genetics* **11**: 449–456.

Jones Jr, H. W. (1997). Record of the first physician to see Henrietta Lacks at the Johns Hopkins Hospital: history of the beginning of the HeLa cell line. *American Journal of Obstetrics and Gynecology* **176**: 227–228.

Jordan, K. and Lynch, M. (1992). The sociology of a genetic engineering technique: Rituals and rationality in the performance of the "plasmid prep." In *The Right Tool for the Job: At Work in Twentieth-Century Life Science*, Clarke, A. and Fujimura, J. (eds.). Princeton, NJ: Princeton University Press, pp. 77–114.

(1998). The dissemination, standardization, and routinization of a molecular biological technique. *Social Studies of Science* **28**: 773–801.

Kaneda, H. (1995). Elimination of paternal mitochondrial DNA in intraspecific crosses during early mouse embryogenesis. *Proceedings of the National Academy of Science USA* **92**: 4542–4546.

Kayser, M., Caglia, A., Corach, D. *et al.* (1997). Evaluation of Y chromosomal STRs: a multicenter study. *International Journal of Legal Medicine* **110**: 125–133, 141–149.

Keller, E. F. (1985). *Reflections on Gender and Science*. New Haven, CT: Yale University Press.

(1986). How gender matters, or, why it's so hard for us to count past two. In *Perspectives on Gender and Science*, Harding, J. (ed.). East Sussex, Philadelphia, PA: The Falmer Press, pp. 168–183.

(1992a). *Secrets of Life, Secrets of Death: Essays on Language, Gender, and Science*. New York: Routledge & Kegan Paul.

(1992b). Nature, nurture, and the Human Genome Project. In *The Code of Codes: Scientific and Social Issues in the Human Genome Project*, Kevles, D. J. and Hood, L. (eds.). Cambridge, MA: Harvard University Press, pp. 281–300.

(1995a). *Refiguring Life: Metaphors of Twentieth-Century Biology*. New York: Colombia University Press.

(1995b). The origin, history, and politics of the subject called "gender and science": a first person account. In *Handbook of Science and Technology Studies*, Jasanoff, S., Markle, G. E., Petersen, J. C., and Pinch, T. (eds.). London: Sage, pp. 95–110.

(1998). Sense and syntax: metaphors of reading in the history of genetics. In *Proceeding of the ASCA Conference, Come to Your Senses*, Amsterdam.

Kevles, D. J. (1985). *In the Name of Eugenics: Genetics and the Issue of Human Heredity*. Cambridge, MA: Harvard University Press.

(1992). Out of eugenics: the historical politics of the human genome. In *The Code of Codes: Scientific and Social Issues in the Human Genome Project*, Kevles, D. J. and Hood, L. (eds.). Cambridge, MA: Harvard University Press, pp. 3–36.

Kevles, D. J. and Hood, L. (eds.) (1992a). *The Code of Codes: Scientific and Social Issues in the Human Genome Project*. Cambridge, MA: Harvard University Press.

(1992b). Reflection. In *The Code of Codes: Scientific and Social Issues in the Human Genome Project*, Kevles, D. J. and Hood, L. (eds.). Cambridge, MA: Harvard University Press, pp. 300–328.

Knorr-Cetina, K. (1981). *The Manufacture of Knowledge: An Essay on the Constructivist and Contextual Nature of Science*. Oxford: Pergamon Press.

(1983). The ethnographic study of scientific work: towards a constructivist interpretation of science. In *Science Observed: Perspectives on the Social Study of Science*, Knorr-Cetina, K. and Mulkay, M. (eds.). London: Sage, pp. 115–140.

(1995). Laboratory studies: the cultural approach to the study of science. In *Handbook of Science and Technology Studies*, Jasanoff, S., Markle, G. E., Petersen, J. C. and Pinch, T. (eds.). London: Sage, pp. 140–166.

Kosto, A. (1994). Besluit DNA Onderzoeken. *Staatsblad van het Koninkrijk de Nederlanden*. Nr. 522.

Krings, M., Stone, A., Schmitz, R. W. *et al.* (1997). Neanderthal DNA sequences and the origin of modern humans. *Cell* **90**: 19–30.

Kuhn, T. S. (1970). *The Structure of Scientific Revolution*, 2nd edn. Chicago, IL: University of Chicago Press.

Laan, M. and Pääbo, S. (1997). Demographic history and linkage disequilibrium in human populations. *Nature Genetics* **17**: 435–438.

Landecker, H. (2000). Immortality, *in vitro*: a history of the HeLa cell line. In *Biotechnology and Culture: Bodies, Anxieties, Ethics*, Brodwin, P. (ed.). Bloomington, IA: Indiana University Press, pp. 53–72.

Lander, E. (1992). DNA fingerprinting: science, law, and the ultimate identifier. In *The Code of Codes: Scientific and Social Issues in the Human Genome Project*, Kevles, D. J. and Hood, L. (eds.). Cambridge, MA: Harvard University Press, pp. 191–210.

Latour, B. (1987). *Science in Action: How to Follow Scientists and Engineers Through Society*. Cambridge: Harvard University Press.

(1988). *The Pasteurization of France*. Cambridge, MA: Harvard University Press.

(1990). Drawing things together. In *Representation in Scientific Practice*, Woolgar, S. and Lynch, M. (eds.). Cambridge, MA: MIT Press, pp. 19–69.

Lauerman J. F. (2001). Arrays cast toxicology in a new light. *NIEHS News*, http://ehpnet1.niehs.nih.gov/docs/2001/109-1/niehsnews.html.

Laurence, D. and Hoekstra, R. F. (1994). Shellfish genes kept in line. *Nature* **368**: 811–812.

Law, J. (ed.) (1991). *A Sociology of Monsters: Power, Technology and the Modern World*. Oxford: Blackwell.

(1994). *Organizing Modernity*. Oxford, Cambridge: Blackwell.

(1999). After ANT: complexity, naming, and topology. In *Actor Network Theory and After*, Law, J. and Hassard, J. (eds.). Oxford: Blackwell, pp. 1–15.

(2000). On the subject of the object: narrative, technology, and interpellation. *Configurations* **8**: 1–29.

(2002). *Aircraft Stories: Decentring the Object in Technoscience*. Durham, NC: Duke University Press.

Law, J. and Callon, M. (1997). The life and death of an aircraft: A network analysis of technical change. In *Shaping Technology/Building Society: Studies in Sociotechnical Change*, Bijker, W. and Law, J. (eds.). Cambridge, MA: MIT Press, pp. 21–53.

Law, J. and Hassard, J. (eds.) (1999). *Actor Network Theory and After*. Oxford: Blackwell.

Law, J. and Williams, R. (1982). Putting facts together: a study of scientific persuasion. *Social Studies of Science* **12**: 535–558.

Lewontin, R. C. (1993). *Biology as Ideology: The Doctrine of DNA*. New York: Harper Perennial.

 (1995a). *Human Diversity*. New York: Scientific American Book.

 (1995b). *DNA Doctrine: Biologie als Ideologie*. Translated by Jos den Bekker. Amsterdam: Uitgeverij Bert Bakker.

Lewontin, R. C. and Hartl, D. L. (1991). Population genetics in forensic DNA typing. *Science* **254**: 1745–1750.

Lewontin, R. C., Rose, S. N. and Kamin L. J. (1984). *Not in Our Genes*. New York: Pantheon.

 (1993). IQ: The rank ordering of the world. In *The "Racial" Economy of Science: Towards a Democratic Future*, Harding, S. (ed.). Bloomington, IA: Indiana University Press, pp. 142–161.

Liloqula, R. (1996). Value of life: saving genes versus saving indigenous people. *Cultural Survival Quarterly*. **20**: Summer. http://www.Culturalsurvival.org/publicationcsq/.

Livingstone, F. B. (1993). On the nonexistance of human races. In *The "Racial" Economy of Science: Towards a Democratic Future*, Harding, S. (ed.). Bloomington, IA: Indiana University Press, pp. 133–142.

Lock, M. (1997). Decentring the natural body: making difference matter. *Configurations* **5**: 267–292.

 (2001). The alienation of body tissue and the biopolitics of immortalized cell lines. *Body and Society* **7**: 63–91.

Lock, M. and Cambrosio, A. (eds.) (2000). *Intersections: Living and Working with the New Medical Technologies*. Cambridge: Cambridge University Press.

Lynch, M. (1985). *Art and Artifact in Laboratory Science: A Study of Shop Work and Shop Talk in a Research Laboratory*. London: Routledge & Kegan Paul.

 (1990). The external retina: selection and mathematisation in the visual documentation of objects in the life sciences. In *Representation in Scientific Practice*, Woolgar, S. and Lynch, M. (eds.). Cambridge, MA: MIT Press, pp. 153–186.

 (1997). *Scientific Practice and Ordinary Action: Ethnomethodology and Social Studies of Science*, 2nd edn. Cambridge: Cambridge University Press.

Lynch, M., Livingston, E. and Garfinkel, H. (1983). Temporal order in laboratory work. In *Science Observed: Perspectives on the Social Study of Science*, Knorr-Cetina, K. and Mulkay, M. (eds.). London: Sage, pp. 205–238.

Macbeth, H. (1993). Ethnicity and human biology. In *Social and Biological Aspects of Ethnicity*, Chapman, M. (ed.). Oxford: Oxford University Press, pp. 47–91.

Macllain, C. (1997). Diversity Project "does not merit" federal funding. *Nature* **389**: 774.

Macpherson, C. B. (1990). *The Political Theory of Possessive Individualism: Hobbes to Locke*, 13th edn. Oxford: Oxford University Press.

Majumder, P. P. (2001). Ethnic populations of India as seen from an evolutionary perspective. *Journal of Bioscience* **26**: 533–545.

Martin, E. (1987). *The Woman in the Body: A Cultural Analysis of Reproduction*. Boston, MA: Beacon Press.

(1991). The egg and the sperm: how science has constructed a romance based on stereotypical male–female roles. *Signs* **16**: 485–501.

(1999). Towards an anthropology of immunology: the body as nation state. In *The Science Studies Reader*, Biagioli M. (ed.). New York: Routledge, pp. 358–372.

Martin, B. and Richards, E. (1995). Scientific knowledge, controversy, and public decision making. In *Handbook of Science and Technology Studies*, Jasanoff, S., Markle, G. E., Petersen, J. C. and Pinch, T. (eds.). London: Sage, pp. 506–531.

Marx, K. (1890; reprinted in 1974). *Das Kapital: Kritik de politischen Ökonomie*, Vol. I. Berlin: Dietz Verlag Berlin.

Marzuki, S., Lertrit, P., Noer, A. S. *et al.* (1992). Reply to Howell *et al.*: the need for a joint effort in the construction of a reference data base for normal sequence variants of human mtDNA. *American Journal for Human Genetics* **50**: 1337–1340.

M'charek, A. (2000). Technologies of population: forensic DNA testing practices and the making of differences and similarities. *Configurations* **8**: 121–158.

M'charek, A. and Rommes, E. (1998). Herkouwende Bewegingen: Een Verslag van een Zomerschool met D. Haraway. *Tijdschrift voor Genderstudies* **3**: 61–66.

Meneva, S. and Ülküer, U. (1995). The distribution of the HLA-DQa alleles and genotypes in the Turkish population as determined by the use of DNA amplification and allele-specific oligonucleotides. *Science and Legal Justice* **35**: 259–262.

Menozzi, P., Cavalli-Sforza, L. L. and Piazza, A. (1994). *The History and Geography of Human Genes*. Princeton, NJ: Princeton University Press.

Metze, M. (1996). *De Staat van Nederland: Op Weg naar 2000. [The "State" of the Netherlands: Towards the Second Millenium.]* Nijmegen: SUN.

Mol, A. (1985). Wie Weet Wat een Vrouw Is? Over de Verschillen en de Verhoudingen Tussen de Disciplines. *Tijdschrift voor Vrouwenstudies* **21**: 10–22.

(1990). Sekse Rijkdom en Bloedarmoede: Over Lokaliseren als Strategie. *Tijdschrift voor Vrouwenstudies* **42**: 142–157.

(1991a). Wombs, pygmentation and pyramids: should anti-racists and feminists try to confine "biology" to its proper place? In *Sharing the Difference*, van Lenning, A. and Hermsen, J. (eds.). London: Routledge, pp. 149–163.

(1991b). Topografie als Methode van Kennisonderzoek: Over het Naast Elkaar Bestaan van Enkele Bloedarmoedes. *Kennis and Methode* **4**: 314–329.

(1998). Dit Geslacht Dat Zoveel Is: Een Conversatie Tussen een Onbekend Aantal Onbekenden van Wie Slechts ÉÉn zich Bekend Zal Maken. *Tijdschrift voor Genderstudies* **1**: 13–15.

(2000). Pathology and the clinic: an ethnography of two atheroscleroses. In *Intersections: Living and Working with the New Medical Technologies*, Lock, M. and Cambrosio, A. (eds.). Cambridge, UK: Cambridge University Press, pp. 82–103.

(2002). *The Body Multiple: Ontology in Medical Practice*. Durham, NC: Duke University Press.

Mol, A. and Hendriks, R. (1995). De Hele Wereld één HB? Universaliteit, Lokaliteit en Bloedarmoede. *Krisis* **58**: 56–73.

Mol, A. and Law, J. (1994). Regions, networks and fluids: anaemia and social topology. *Social Studies of Science* **24**: 641–671.

Mol, A. and Mesman, J. (1994). Neonatal food and the politics of theory: some questions of method. *Social Studies of Science* **26**: 419–444.

Molnar, S. (1975). *Races, Types, and Ethnic Groups: The Problem of Human Variation.* Englewood Cliffs, NJ: Prentice-Hall.

Moser, I. and Law, J. (1999). Good passages, bad passages. In *Actor Network Theory and After*, Law, J. and Hassard, J. (eds.). Oxford: Blackwell, pp. 196–219.

NRC (1992). *DNA Technology in Forensic Science.* Washington, DC: National Academy Press for the National Research Council.

(1996). *The Evaluation of Forensic DNA Evidence.* Washington, DC: National Academy Press for the National Research Council.

(1997). *Evaluating Human Genetic Diversity.* Washington, DC: National Academy Press for the National Research Council.

Novas, C. and Rose, N. (2000). Genetic risk and the birth of the somatic individual. *Economy and Society* **29**: 485–513.

Olivo, P. D., van de Walle, M. J., Laipis, P. J. and Hauswirth, W. W. (1983). Nucleotide sequence evidence for rapid genotypic shift in the bovine mitochondrial DNA D-loop. *Nature* **306**: 400–402.

Olson, S. (2001). The genetic archaeology of race. *The Atlantic Monthly.* April: 69–80. http://www.theatlantic.com/issues/2001/04/olson-p1.htm.

Oudshoorn, N. (1994). *Beyond the Natural Body: An Archaeology of Sex Hormones.* London: Routledge.

Pääbo, S. (1995). The Y-chromosome and the origin of all us (men). *Science* **268**: 1141–1142.

(1996). Mutation hot spots in the mitochondrial microcosm. *American Journal of Human Genetics* **59**: 493–496.

Parkin, R. (1997). *Kinship: An Introduction to the Basic Concepts.* Oxford: Blackwell.

Pasveer, B. (1992). Shadows of knowledge: making a representing practice in medicine: X-ray pictures and pulmonary tuberculosis, 1895–1930. Ph.D. Thesis, University of Amsterdam.

Pasveer, B. and Akrich, M. (1998). Hoe Lichamen Circuleren: Over Definities van het Zwangere Lichaam, Medische technologie en de Toekomst van de Thuisbevalling. *Tijschrift voor Genderstudies* **3**: 47–56.

Pickering, A. (ed.) (1992). *Science as Practice and Culture.* Chicago, IL: University of Chicago Press.

Rabinow, P. (1993). Galton's regret: of types and individuals. *Culture, Medicine and Psychiatry* **17**: 59–65.

(1996a). *Making PCR: A Story of Biotechnology.* Chicago, IL: University of Chicago Press.

(1996b). *Essays on the Anthropology of Reason.* Princeton, NJ: Princeton University Press.

RAFI (1993). *Patents, Indigenous People, and Human Genetic Diversity.* Ottawa: RAFI Communique, May 1993.

Rapp, R. (2000). *Testing Women, Testing the Fetus: The Social Impact of Amniocentesis in America.* New York: Routledge.

Reardon, J. (2001). The Human Genome Diversity Project: a case study in coproduction. *Social Studies of Science* **31**: 357–388.

Reid, R. and Traweek, S. (eds.) (2000). *Doing Science + Culture: How Cultural and Interdisciplinary Studies are Changing the Way we Look at Science and Medicines.* New York: Routledge.

Rheinberger, H.-J. (1997). Von der Zelle zum Gen: Repräsentationen der Molekularbiologie. In *Räume des Wissens: Repräsentation, Codierung, Spur*, Rheinberger, H.-J., Hagner, M. and Wahrig-Schmidt, B. (eds.). Berlin: Akademie Verlag, pp. 265–279.

(1999). Experimental systems: historiality, narration, and deconstruction. In *The Science Studies Reader*, Biagioli, M. (ed.). New York: Routledge, pp. 417–429.

(2001). *Experimentalsysteme und Epistemische Dinge: Eine Geschichte der Proteinsynthese im Reagenzglas.* Göttingen: Wallstein.

Roberts, L. (1991). A genetic survey of vanishing peoples. *Science* **252**: 1614–1617.

Roewer, L., Arnemann, J., Spurr, N. K., Grzeschik, K. H. and Epplen, J. T. (1992). Simple repeat sequences on the human Y chromosome are equally polymorphic as their autosomal counterparts. *Human Genetics* **89**: 389–394.

Roewer, L., Kayser, M., Dieltjes, P. *et al.* (1996). Analysis of molecular variance (AMOVA) of Y-chromosome-specific microsatellites in two closely related human populations. *Human Molecular Genetics* **5**: 1029–1033.

Rubin, G. (1975). The traffic in women: notes on the political economy of sex. In *Toward an Anthropology of Women*, Rapp Reiter, R. (ed.). New York: Monthly Review Press, pp. 157–210.

Rubinzstein, D. C., Leggo, J. and Amos, W. (1995) "Microsatellite evolution: evidence for directionality and variation in rate between species." *Nature Genetics* **10**: 337–341.

Sajantila, A., Lahermo, P., Anttinen, T. *et al.* (1995). Genes and languages in Europe: an analysis of mitochondrial lineages. *Genome Research* **5**: 42–52.

Sajantila, A., Salem, A.-H., Savolainen, P., *et al.* (1996). Paternal and maternal DNA lineages reveals a bottleneck in the founding of the Finnish population. *Proceedings of the National Academy of Science USA* **93**: 12035–12039.

Salem, A., Badr, F. M., Gaballah, M. F. and Pääbo, S. (1996). The genetics of traditional living: Y-chromosomal and mitochondrial lineages in the Sinai peninsula. *American Journal of Human Genetics* **59**: 741–743.

Sanger, F. and Coulson, A. R. (1975). A rapid method for determining sequences in DNA by primed synthesis with DNA polymerase. *Journal of Molecular Biology* **94**: 441–448.

Sanger, F., Nicklen, S. and Coulson, A. R. (1977). DNA sequencing with chain-terminating inhibitors. *Proceedings of the National Academy of Science USA* **74**: 5463–5467.

Sanker, P. and Cho, M. K. (2002). Towards a new vocabulary of human genetic variation. *Science* **298**: 1337–1338.

Schull, W. J. (1997). Correspondence. *Nature* **390**: 221.

Singleton, V. (1998). Stabilizing instabilities: the role of the laboratory in the United Kingdom cervical screening programme. In *Differences in Medicine: Unravelling Practices, Techniques and Bodies*, Berg, M. and Mol, A. (eds.) Durham, NC: Duke University Press, pp. 86–105.

Sjerps, M. and Kloosterman, A. D. (1999). On the consequences of DNA profile mismatches for close relatives of an excluded suspect. *International Journal of Legal Medicine* **112**: 176–180.

Skibinski, D. O., Gallagher, C. and Beynon, C. M. (1994).'Mitochondrial DNA inheritance. *Nature* **368**: 817–818.

Smith, J. M. and Smith, N. H. (1998). Detecting recombination from gene trees. *Molecular Biology and Evolution* **15**: 590–599.

(2002). Recombination in animal mitochondrial DNA. *Molecular Biology and Evolution* **19**: 2330–2332.

Smith, S. J. (1993). Residential segregation and the politics of racialization. In *Racism, the City and the State*, Cross, M. and Keith, M. (eds.). New York: Routledge, pp. 128–143.

Stamhuis, I. H. and de Knecht-van-Eekelen, A. (eds.) (1997). Zij is toch wel zeer begaafd: Historische bijdragen over vrouwen in de bètawetenschappen. *Gewina* **4**: special issue.

Star, S. L. (1991). Power, technology and the phenomenology of conventions: On being allergic to onions. In *A Sociology of Monsters: Essays on Power, Technology and Domination*, Law, J. (ed.). Oxford: Blackwell, pp. 26–56.

(1995a). The politics of formal representations: wizards, gurus, and organizational complexity. In *Ecologies of Knowledge: Work and Politics in Science and Technology*, Star, S. L. (ed.). New York: State University of New York Press, pp. 88–118.

(ed.) (1995b). *Ecologies of Knowledge: Work and Politics in Science and Technology*. Albany, NY: State University of New York Press.

(1995c). Introduction. In *Ecologies of Knowledge: Work and Politics in Science and Technology*, Star, S. L. (ed.). New York: State University of New York Press, pp. 1–35.

Stepan, N. L. (1986). Race and gender: the role of analogy in science. *Isis* **77**: 261–277.

Stepan, N. L. and Gilman, S. (1993). Appropriating the idioms of science: the rejection of scientific racism. In *The "Racial" Economy of Science: Towards a Democratic Future*, Harding, S. (ed.). Bloomington, IA: Indiana University Press, pp. 170–193.

Strathern, M. (1988). *The Gender of the Gift: Problems with Women and Problems with Society in Melanesia*. Berkeley, CA: University of California Press.

(1991). *Partial Connections*. Savage, MD: Rowman & Littlefield.

(1992). *Reproducing The Future: Anthropology, Kinship and the New Reproductive Technologies*. Manchester: Manchester University Press.

(1995a). *After Nature: English Kinship in the Late Twentieth Century*. Cambridge: Cambridge University Press.

(1995b). The nice thing about culture is that everybody has it. In *Shifting Contexts: Transformations in Anthropological Knowledge*, Strathern, M. (ed.). London: Routledge, pp. 153–177.

(1995c). Displacing knowledge: technology and the consequences for kinship. In *Conceiving the New World Order: The Global Politics of Reproduction*, Ginsburg, F. D. and Rapp, R. (eds.). Berkeley, CA: University of California Press, pp. 346–364.

Sykes, B. (2001). *The Seven Daughters of Eve*. London: Corgi Books.

Te Pareake Mead, A. (1996). Genealogy, sacredness, and the commodity market. *Cultural Survival Quarterly* **20**: 46–51. http://www.Culturalsurvival.org/publicationcsq/.

Thain, M. and Hickman, M. (eds.) (1996). *Penguin Dictionary of Biology*. London: Penguin Books.

Thorne, A. G. and Wolpoff, M. H. (1992). The multiregional evolution of humans. *Scientific American* **266**: 28–33.

Timmermans, S. and Berg, M. (1997). Standardisation in action: achieving local universality through medical protocols. *Social Studies of Science* **27**: 237–305.

Titmuss, R. M. (1997). *The Gift Relationship: From Human Blood to Social Policy*, 2nd edn. London: LSE Books.

Torroni, A. and Wallace, D. C. (1995). mtDNA haplogroups in Native Americans. *American Journal of Human Genetics* **56**: 1234–1236.

Traweek, S. (1988). *Beamtimes and Lifetimes: The World of High Energy Physicists*. Cambridge, MA: Harvard University Press.

(1992). Border crossings: Narrative strategies in science studies and among physicists in Tshukuba science city, Japan. In *Science as Practice and Culture*, Pickering, A. (ed.). Chicago, IL: University of Chicago Press, pp. 429–465.

(1999). Pilgrim's progress: male tales told during a life in physics. In *The Science Studies Reader*, Biagioli, M. (ed.). New York: Routledge, pp. 525–542.

Tutton, R. (1998). Culture and identity in European genetic diversity. In *Proceedings of the PFGS Colloquium 2*, University College London, December 1998.

UNESCO (1951). *UNESCO and its Programme III: The Race Question*. [Publication 785.] Paris: UNESCO.

(1952). The race concept: results of an inquiry. In UNESCO. *The Race Question in Modern Science*. Paris: UNESCO, pp. 36–91.

van Kampen, P. T. C. (1998). *Expert Evidence Compared: Rules and Practices in the Dutch and American Justice System*. Antwerp: Intersentia Rechtswetenschappen.

Verran, H. (2002). A postcolonial moment in science studies: alternative firing regimes of environmental scientists and aboriginal landowners. *Social Studies of Science* **32**: 729–762.

von Haeseler, A., Sajantila, A. and Pääbo, S. (1996). The genetic archaeology of the human genome. *Nature Genetics* **14**: 135–140.

Wade, P. (2002). *Race, Nature and Culture: An Anthropological Perspective*. London: Pluto Press.

Wakely, J. (1993). Substitution rate variation among sites in hypervariable region I of human mitochondrial DNA. *Journal of Molecular Evolution* **37**: 613–623.

Wallace, R. W. (1998). The Human Genome Diversity Project: medical benefits versus ethical concerns. *Molecular Medicine Today* **4**: 59–62.

Watkins, W. S. Bamshad, M. and Jorde, L. B. (1995). Population genetics of trinucleotide repeat polymorphisms. *Human Molecular Genetics* **4**: 1485–1491.

Watson, J. D., Gilman, M., Witkowski, J. and Zoller, M. (eds.) (1991). *Recombinant DNA*, 2nd edn. New York: Scientific American Books.

Watson-Verran, H. and Turnbull, D. (1995). Science and other indigenous knowledge systems. In *Handbook of Science and Technology Studies*, Jasanoff, S., Markle, G. E., Petersen, J. C. and Pinch, T. (eds.). London: Sage, pp. 115–140.

Wills, C. (1996). Another nail in the coffin of the multiple-origins theory? *Bioessays* **18**: 1017–1020.

Wilson, A. C. and Cann, R. (1992). The recent African genesis of humans. *Scientific American* **266**: 22–27.

Wilson, A. C., Cann, R. L., Carr, S. M. *et al.* (1985). Mitochondrial DNA and two perspectives on evolutionary genetics. *Biological Journal of the Linnean Society* **26**: 375–400.

Woolgar, S. and Latour, B. (1986). *Laboratory Life: The Social Construction of Scientific Facts.* Beverly Hills, CA: Sage.

Yang, F., O'Brien, P, C., Milne, B. S. *et al.* (1999). A complete comparative chromosome map for the dog, red fox, and human and its integration with canine genetic map. *Genomics* **62**: 182–202.

Zouros, E. (1994a). 'Mitochondrial DNA inheritance. *Nature* **368**: 817–818.

(1994b). An unusual type of mitochondrial DNA inheritance in the blue mussel *Mytilus. Proceedings of the National Academy of Science USA* **91**: 7463–7467.

Index